Synthesis Lectures on Information Concepts, Retrieval, and Services

Series Editor

Gary Marchionini, School of Information and Library Science, The University of North Carolina at Chapel Hill, Chapel Hill, NC, USA

This series publishes short books on topics pertaining to information science and applications of technology to information discovery, production, distribution, and management. Potential topics include: data models, indexing theory and algorithms, classification, information architecture, information economics, privacy and identity, scholarly communication, bibliometrics and webometrics, personal information management, human information behavior, digital libraries, archives and preservation, cultural informatics, information retrieval evaluation, data fusion, relevance feedback, recommendation systems, question answering, natural language processing for retrieval, text summarization, multimedia retrieval, multilingual retrieval, and exploratory search.

Laurie J. Bonnici · Brian C. O'Connor

Proximity and Epidata

Attributes and Meaning Modification

 Springer

Laurie J. Bonnici
College of Communication and Information
Sciences
The University of Alabama
Tuscaloosa, AL, USA

Brian C. O'Connor
Department of Information Science, College
of Information
University of North Texas
Denton, TX, USA

ISSN 1947-945X   ISSN 1947-9468 (electronic)
Synthesis Lectures on Information Concepts, Retrieval, and Services
ISBN 978-3-031-17096-6   ISBN 978-3-031-17094-2 (eBook)
https://doi.org/10.1007/978-3-031-17094-2

This Springer imprint is published by the registered company Springer Nature Switzerland AG
The registered company address is: Gewerbestrasse 11, 6330 Cham, Switzerland

Thinking of Elfreda Chatman
Elfreda, the intellectual banter continues…

Laurie J. Bonnici

Thinking of Patrick Wilson
PW, can one really say that?

Brian C. O'Connor

Contents

About the Authors

 **Co-authors Laurie J. Bonnici and Brian C. O'Connor** are professors of information science, University of Alabama and University of North Texas, respectively. They are academic descendants of pioneering information philosopher, Patrick Wilson of U. C. Berkeley, School of Library and Information Studies. Their work weaves O'Connor's visual information approaches with Bonnici's language analysis to examine the information at the edges. They enjoy kayaking and exploring coffee houses while navigating wherever they may be. Toronto provided fertile ground for conceiving the notion of anecdata and proximity in information experiences.

List of Figures

Proximity and Clues

1

1.1   Introduction

Our interest in proximity began two decades ago when a mutual friend introduced us because of our common passion for kayaking. As we chatted about types of kayaks and paddles, our conversation drifted to our work. We each began to think about some of the other's comments: "I've heard this before." The simple question "Who was your major professor?" yielded our first discussion of proximity. Brian had studied under Patrick Wilson, Laurie had studied under Elfreda Chatman, who had studied under Wilson at the same time Brian had—we had a close intellectual relationship, a proximity that had been unknown until that moment.

Over the years, physical proximity has been irregular and less than frequent, yet we have engaged in various projects and critiqued each other's work based on our conceptual proximity, aided by the tele-proximity of email, telephone, Skype, and Zoom. Recently, we had the opportunity to spend some time together to work out a presentation for the Document Academy (DOCAM) on professors as dynamic documents. Our close but differing ties to Wilson surfaced in our discussions, largely in terms of the continuity of that dynamic quality through students and students of students—the roles of memory and critical consideration through generations. Kayaks also came into play. Kayaks have little or nothing to do with Patrick Wilson, but our passion for kayaking was the thread that brought us together. Would we have met and had a conversation that brought up the same point? It is possible, but not necessarily the likely case. This led to our thinking about the role of unexpected, unknown, unpredicted threads of proximity—the kayaks might be seen as simple catalysts in retrospect, yet they were a strong incentive to have the first meeting and provided an affordance for both conversations and metaphors for our thinking and doing. Such unexpected connections are worth examining as components of information seeking and retrospective analysis.

© The Author(s), under exclusive license to Springer Nature Switzerland AG 2022   1
L. J. Bonnici and B. C. O'Connor, *Proximity and Epidata*,
Synthesis Lectures on Information Concepts, Retrieval, and Services,
https://doi.org/10.1007/978-3-031-17094-2_1

We set out to sketch a constellation of proximities, present examples of attempts to accomplish proximity, and provoke a discussion of the role of proximity in our field. We suggest that proximity is a thread between retrieval constructs based on known topics and predictable relations and the sorts of information seeking that lie outside those constructs—browsing, stumbling, encountering, detective work, art making, translation, understanding documents in hand.

We see our work here more as a set of provocations than as a single, complete model—a meandering which "[may seem] to be a slower process than the straight line of progress, [but is only so] for the simply defined objective, for the short view of time; for meandering proceeds by covering more ground, percolating into deeper depths, listening to the murmurs of more voices, being what it is when and where it is observed." (Klaver, 2018) We cover quite a bit of ground, paddle some lakes and rivers, and invite the murmurs of more voices.

1.2   Once Upon a Time

Our story of Proximity begins with an ancient Greek legend. Theseus had a problem and he had no clue how to solve it.

For some time, the people of Athens had been compelled to send young men and women to Knossos to be sacrificed to the Minotaur—a half human, half bull creature born to the wife of King Minos as punishment for Minos insulting the god Poseidon. Theseus, being of the right age, promised his father King Aegeus he would slay the creature and free his people of this horrible burden.

Killing the Minotaur would not be easy; yet, as an accomplished warrior, Theseus was confident he could slay the creature, as shown in Fig. 1.1 (Diosphos, 500 BCE). The real problem was what to do after the killing. The Minotaur was enclosed in a labyrinth, as visualized in Fig. 1.2 (van de Passe, 1602) designed by the legendary designer and maker, Daedalus, to be inescapable—keeping the dangerous creature and his sacrificial meals trapped together and away from the citizenry of Knossos. Escaping the inescapable labyrinth was the real problem for which Theseus had no clue.

Enter Ariadne, daughter of King Minos. As often happens in such tales, Ariadne fell in love with Theseus, so she was invested in his problem of escaping the labyrinth. She went to Daedalus and asked for help. He suggested thread, so Theseus would have a physical connection to the outside—maintaining proximity to Ariadne. At the appointed hour Theseus prepared his sword and Ariadne gave him a ball of thread—literally a "clew or clue," a skein of yarn or thread. Theseus entered the maze letting out the line from his clew, slew the creature, and wound/found his way back out through the otherwise inescapable labyrinth. Ariadne knew Daedalus and Theseus knew Ariadne; the chain of proximity yielded a clue for a useful solution to a vexing problem.

Fig. 1.1 Theseus slays the Minotaur. Metropolitan Museum of Art

Fig. 1.2 Labyrinth with Minotaur and Theseus, portion of etching by Crispijn van de Passe (1), 1602–1607. Rijksmuseum, Amsterdam

We cannot always count on Ariadne going to Daedalus to help us solve problems, but humans do hunt and gather useful clues in numerous ways. It is our aim to set out examples of proximity and clues in order to suggest an inclusive model of information seeking and use. We hope, also, to provoke a broader conversation.

1.3   Clues, Proximity, and Functionality

It is useful to turn to the etymology of "clue" (a variant of the earlier "clew"). The *Oxford English Dictionary* (Weiner & Simpson, 2004) tells us that a clue is: "A ball of thread, which in various mythological or legendary narratives (esp. that of Theseus in the Cretan Labyrinth) is mentioned as the means of 'threading' a way through a labyrinth or maze; hence, in many more or less figurative applications: that which guides through a maze, perplexity, difficulty, intricate investigation, etc."

The *OED* gives as it earliest example of this meaning of "clue" as Chaucer's *Legend Good Women* Ariadne from 1385: By a clewe of twyn as he hath gon The same weye he may returne a-non ffolwynge alwey the thred as he hath come.

The *OED* also alerts us that, while physical closeness is the usual meaning of "proximity," more abstract senses can also be implied: (1) The fact or condition of being near or close in abstract relations, as kinship, time, nature, etc.; closeness; (2) The fact, condition, or position of being near or close by in space; nearness. Now the dominant sense.

The tale of Theseus and Ariadne presents us with different facets of proximity. Theseus has to sail from Athens to Knossos just to be proximate with the Minotaur. Ariadne becomes close to Theseus and wanting to maintain physical and emotional proximity to him decides to help solve the problem of exiting the labyrinth. Neither Ariadne nor Theseus has any idea of how to escape the labyrinth; however, Ariadne is close to Daedalus, designer of the labyrinth. It is Daedalus who comes up with the idea of using thread to establish a connection to the exit—a clue, a route marker for maintaining proximity to the exit. Had Theseus not met Ariadne and had she not known someone who might have an idea, the Minotaur would no longer have been a threat to Athens, but Theseus would not have returned home.

We assume documents are a means of proximity to what was in the minds (broadly speaking) of their authors. However, finding and understanding documents is another layer of proximity between the coding practices of authors and the decoding abilities users. In general, documents are not sought simply for their physical presence; rather they are sought for what they enable us to do, for the states into which they might enable us to put our neural pathways, whether to solve a unique problem, engage in a lovely memory, remember what to buy at the grocery, find reviews of blood pressure cuffs, or …

Grocery lists, vacation photos, dissertations, journal articles are attempts to put the reader/viewer/user into proximity with the author. Grocery lists and vacation photos may be seen by nobody other than the author, yet even in this case the author is put back in time and, perhaps place—standing before the empty shelf in the pantry two hours earlier, or back at a fjord in Norway in 1973. Damasio asserts that remembering something is putting neural pathways into the same state they were when the target of remembering originally happened. (Damasio) The documents act as clues, codes to help neural pathways become proximate to some earlier state. In a conversation, Brian asked Damasio if a novelist or film director could be said to be putting readers and viewers into neural states they had never actually experienced; Damasio replied that he would not argue against that idea. (MIT Image & Meaning Symposium, 2005. Personal communication.)

We assume meaning is (largely) function. Authors generate documents in order to share some functionality—perhaps only with themselves and those close to them or with some wider group; seekers want some functionality. The documents (messages) generated and sought are of all sorts—formally coded books, sculptures, movies, symphonies, videos of the cat, having beer and pizza with a professor while working out a dissertation idea, photographs of knots for holding ribs in a kayak, reading Ginsberg's *HOWL*, hearing Ginsberg perform *HOWL*.

We use the term epidata as a general term for clues to finding and understanding useful information. We should note that epidata includes dynamic and mutable constructs. Theseus was the first to think of entering a labyrinth, killing a menacing creature, then exiting the labyrinth. There was no context, no established knowledge on how to proceed. Thread as a means of escaping a labyrinth could not have been found in Theseus's local library; yet, now that we know the solution worked, we can move it into a standard index term.

Wilson speaks of moving away from the topical or disciplinary organization and description of documents and instituting a "reorientation toward the functional." (Wilson, 1977). Then he notes such an approach "explicitly recognizes the primacy of the need to bring knowledge to the point of use." Now, there is a problem of a sort here. Much of what has been done in the past has been a matter of labeling—labeling what the primary subject of a document is and labeling what the primary subject of a question is. This makes it possible to match a user's question to a document—essentially resulting in an efficient warehousing system. However, this assumes that someone seeking information actually knows just how to state what it is that is troubling, missing, vexing, or needed. It also assumes that someone knows how to translate whatever is in a retrieved document to their problem. It also assumes there is a document available. It also assumes that the document and the seeker employ a common language in a mutually expressed and understood way. Wilson (his works are just full of insights stated in wonderful prose) notes that if one "made knowledge available for use by making documents available for use, the librarian's role would be clear-cut, [b]ut the availability of knowledge is not like the availability of an object – a book, a hammer, or a loaf of bread."

1.4   Epidata

The difference between seeing and understanding lies in knowing the context, and it should be emphasized that we approached this problem area from a particular perspective: helping readers to understand.

Michael Buckland and Michele Renee Ramos.

Events as a Structuring Device in Biographical Mark-up and Metadata.

Bulletin of the American Society for Information Science and Technology. December/January 2010.

Our story of proximity turns to a food court in Toronto. For some time, we had been thinking about professors (among others) as dynamic documents, digesting new information, critiques, and insights and updating their concepts. Having ideas rolling around in their head that they are yet processing. Those thoughts and critiques can be delivered to colleagues and students as soon as they are formulated and within the context at hand. For students looking a little baffled a simple example and some extra background material can be provided for clarification of understanding. As for colleagues in the field, much may be bypassed to get to further discussion of the new ideas (Fig. 1.3).

We had been considering the concept of "professor as document" for some time largely attempting to model increasing constraints on the professoriate from many sides. While chatting and scribbling notes in a food court we turned our thoughts to Patrick Wilson as a professor of note to whom we each had strong but quite different connections—proximity. We reasoned it could help readers of Wilson's philosophical works to know of his hardy laugh by making him appear more human and thus approachable. In the first two pages

Fig. 1.3 First notes on proximity and anecdata

of his oral history (Wilson & McCreery, 2000) the note "[laughter]" appears three times; Howard White noted from his doctoral student days:

> Wilson was a good lecturer—earnest but also playful in a style … with a crooked smile and the slightly odd habit of sometimes grasping his ankle and resting his knee on the front table, which gave him a crane-like stance. (White, 2000)

It might be useful in some way for them to know that his passion and rigor was so deep that even after he had retired, he had a dream that he might have made an errant perception about positivism, and that he then went back through his writings to be sure his arguments still held. Brian then recalled an incident involving Wilson's first book.

In 1997, Brian had assigned Wilson's *Two Kinds of Power* to a group of master's degree students in his course. The text was just as it had been when set in print thirty years earlier; no new versions, update, or changes had been made. Most of the students declared they would not read the book because Wilson was "too sexist"—he did not use gender-neutral pronouns and he used masculine phrases such as "How is a man to know…?" The writing style followed the norms of publishing for the time Wilson published the work—1968. Changes in cultural attitudes on gender over the ensuing thirty years combined with more static norms of publishing resulted in a breakdown in the functionality of Wilson's text to connect with the students. This was an example of how small differences in circumstances might go unnoticed, yet cause significant problems distancing a possible user and a source of potentially useful information. Common ground was not so substantial as it seemed—there was a lack of temporal proximity. When Wilson wrote the book there was no style manual that enforced gender-neutral pronouns. Brian posited to the students that if anyone would have argued for gender neutral use in publishing, Wilson would likely have been among them (from personal conversations with him). Even if he had used them, the publisher would have "edited" them away. Sharing these perspectives on Wilson with the students opened a pathway of proximity that shifted students' perspective. This shift led to their reading the work with (self-reported) enthusiasm. Here we see that Brian provided information around the book (epidata) that led to a connection with the original and unchanged information (book).

Laurie noted that such anecdotes might be helpful in translating across all manner of disruptions of functionality caused by lack of proximity to an information source. The very word "anecdote" prompted our original and early construct anecdata. We originally constructed "anecdata" to resonate with metadata, yet be significantly different. The term "anecdote" comes from roots meaning "unpublished"—this does not mean it is not evidentiary or of little use, only that anecdata is not 'publicly" available in any ordinary sense. The Oxford English Dictionary notes:

> **anecdote (n.)**1670s, "secret or private stories," from French *anecdote* (17c.) or directly from Medieval Latin *anecdota*, from Greek *anekdota* "things unpublished," neuter plural of *anek-dotos*, from *an-* "not" + *ekdotos* "published"

Procopius' 6c. *Anecdota*, unpublished memoirs of Emperor Justinian full of court gossip, gave the word a sense of "revelation of secrets," which decayed in English to "brief, amusing story" (1761).

As the construct of anecdata evolved, we noted a lack of depth in its fully expressing connection with information. Thus, we have coined the term epidata to reflect our idea of information that surrounds information; a term borrowed and adjusted from epigenetics.

- Epigenetics—the study of changes in organisms caused by modification of gene expression rather than alteration of the genetic code itself.
- Epidata—the changes in content around or accompanying information that modifies meaning or understanding rather than alteration of the information itself.

In our musings about professors as documents, we might consider epidata as a form of continuous development of faculty as document (connective synchronicity), even in the posthumous state. Little known knowledge about the distant and departed can continue to open doors for the diachronic documents left behind. Proximity to an author may enhance understanding of that author's works, yet not all readers are close to an author. We propose epidata as a means of achieving virtual proximity.

Proximity comes from Latin roots meaning "in the vicinity"—for us it is an understanding of an author afforded through closeness ties—closer proximity yields higher probability of informative epidata. For example, a doctoral student has proximity to a major professor; a student of that student would have virtual proximity, depending on what was remembered and passed on; those who read something such as White's words on Wilson might have a third- generation virtual proximity; whereas a student with only the book would have only a reading familiarity (weak tie) and would lack the embellishing perspective epidata that enriches content.

1.5   Diachronicity and Synchronicity—Epidata (Connective Synchronicity)

Documents are characterized by both static and dynamic attributes allowing for their function and relevance, and even their potential reduction of use over time. O'Connor provides a concise overview of this phenomenon. Through notions of diachronicity and synchronicity he explains how documents retain and lose value as useful information (O'Connor et al., 2008, p. 41). In his presentation, synchronicity is characterized by societal and cultural changes in meaning over time. He posits that such changes, depending on context, allow for growth of use or death of information. Recognizing the diachronic characteristics of documents, here we build on the notion of synchronicity as we conceptualize epidata.

We can say that a document (main source of information) is a message together with all the relations surrounding that message (epidata): a set of words, pictures, sounds together with what physical and social constraints urged the author to choose and position certain signs in just such a way and what caused (enabled, constrained) any recipient to find the message useful. Barring decay of media, messages remain stable across time, while the mix of antecedents to making a message and the complementary mix of antecedents to understanding coding and decoding) are situational.

In the 1960s people destroyed rock and roll records, now the same messages can be heard as background music at the supermarket. Musician's lives evolved including one of the Beatles was bestowed with a knighthood and Brian May (Queen) earned a doctorate in astrophysics. The message (music) persisted through time (diachronic) while some of the attributes of the artists changed (synchronic); thus, we might say, the document changed. The ties between the state of affairs at play when the message was made and those at play for any particular author/recipient can be termed epidata.

It is easy enough to recognize a decoding problem when there is a significant difference between the message maker and a recipient. μῆνιν ἄειδε θεὰ Πηληϊάδεω Ἀχιλῆος οὐλομένην is recognizably in a different coding system from that of the rest of this text. A photo may make little sense without context. Where epidata become useful are in those circumstances where coding/decoding differences are seemingly slight yet stand in the way of discovery and understanding.

The Four Tops express quite clearly the diachronic/synchronic dance with their lyrics:

You're sweet as a honey bee

But like a honey bee stings

You've gone and left my heart in pain

All you left is our favorite song

The one we danced to all night long

It used to bring sweet memories

Of a tender love that used to be

Now it's the same old song

But with a different meaning

Since you been gone

It's the same old song

But with a different meaning

Since you been gone

I, oh I (Holland et al., 1965)

We propose proximity and epidata to account for ways of finding and understanding functional information, ways that stand outside yet in concert with the norms of data and metadata. Epidata afford pathways, point to details that cast light on proximities that might otherwise go unknown. Epidata are those relationships not readily available or known from a document or the metadata around it. Epidata may be in the memory of a friend of an author, in knowledge of the culture within which a document was constructed; they may be immediately available upon looking in the right place or they may require significant detective work for discovery. Epidata bring one closer to the author of a message, thereby aiding discovery, interpretation, and understanding of useful documents. Epidata make up the clue, the ball of thread that affords proximity while leaving the original information, in its original presentation, unchanged.

Data, metadata, and epidata are all anchor points for clues—the threads between where one is and where one wants to be—between oneself and that to which one wants to be proximate. Some of those anchor points are at (for our purposes) the document end and some of those anchor points are at the seeker's end. It would have done little good for Ariadne to stand at the entrance to the Labyrinth and throw the clue randomly into a passageway—it had to be attached to Theseus for him to be able to again get close to Ariadne. We might say in general that data rest largely at the document side, metadata act as general descriptions of typical pathways between documents and seekers, and epidata rest at the seeker side.

1.6   Epidata: Clues to Proximity

There are many ways to find information about things we know we need to know. Yet, finding the most suitable information is not so simple a task. What of those times when something "isn't quite right,"; what about times when what we have found does not make sense, even though it "should;" consider those times when we are not looking for information at all, but something "catches our eye?"

A document (or document event) is co-constituted by a message maker & a message recipient—both impacted by antecedents—cultural and personal constraints. Looking at the clues that can bring users of documents into a closer relation with the authors of the documents is the form of proximity we are examining. When somebody visits a place, builds something, or solves a problem, we say they have first-hand knowledge of that experience; when that person shows photographs, draws up plans, or prints an equation, we say that they have created a document to provide second-hand knowledge of the event. The closer the author and user are in terms of language, assumptions, life experiences, the more likely it is that the user will more fully understand the document. So we can say that author/user proximity is a way of understanding second-hand knowledge as if it were first-hand knowledge.

Imagine being able to walk about the agora with Plato and observe the questioning style. Imagine observing the whirlwind five-day writing spree on the film version of *Gone with the Wind*. Imagine understanding why people in nineteenth century photos seldom smile; understanding why Weston's Pepper #30 is so astonishing; understanding why a two-line poem by Catullus was once shocking (and why Carl Orff's version is so very unlike that of Catullus). Consider the "what if" questions and responses while having coffee with Zenodotus and Tim Berners Lee? Getting the subtle details around, but not central to, the information may be the catalyst to an aha moment information encounter.

What of those times when we just wonder what someone from our hometown would think of this or that? What if we see that a current best-selling book on a topic for which we have little interest was authored by a college classmate—we might pick it up. What if we are a filmmaker between projects and happen to see a position open for a film archivist, see fine print "requires MLS," and we enroll in graduate school? What if we are a scholar looking for inspiration? We might put ourselves in the library stacks looking at documents with which we are not familiar, in hopes of stumbling upon something useful? What if we want people to know the music of some of the blues greats, but we know we are not so talented as they are? We might play the music on loud electric instruments, speed the music up, call ourselves the Rolling Stones, and have our idols thank us for our efforts. All of these hint at the uses of proximity and the necessity for knowing the seemingly little hooks that make finding and understanding of information possible.

When Patrick Wilson called for "…a reorientation toward the functional" and "the need to bring information to the point of use" the information world was on the cusp of the Internet age (Wilson, 1977). At that time, computers were used for some limited information organization tasks. For the most part, documents largely remained paper-based. The joy of instantaneous access to a veritable sea of digital information made available through a search engine unwittingly circumnavigated human intermediaries; individuals holding potential to bridge author and reader that have the potential of opening the possibilities for new ways of discovering connections with useful information.

Creative and scholarly works are a part of the author, their personal experiences potentially emboldening content. The static nature of documents presented as books, journals, and even websites, eludes the dynamic representation embodied in the author. Anecdotal fodder offers small differences in common ground between reader and the document opening a window of potential interest. The subtle connection holds potential to bring what may have been unnoticed information to the point of use.

We have envisioned the concept of epidata as a connection between reader/viewer/listener and document, where anecdotal information about the author breathes new life, new functionality into the document. What are those attributes, those characteristics of an author or a work or of a reader/user/viewer/listener that might be helpful but are not now routinely gathered in some sense? Epidata are the data around or accompanying information that modifies meaning or understanding rather than alteration of the information itself. What are the sorts of characteristics of the relationship between an author and document and user that might be useful that are, perhaps, unknowable beforehand, but for which one might train/advise a seeker to consider in any particular instance?

Let us set about our meandering.

References

Damasio, A. (2012). *Self comes to mind: Constructing the conscious brain.* Vintage Books.
Diosphos Painter (ca. 500 BCE). Metropolitan Museum of Art, gift of Dietrich von Bothmer, 1964. Available through The Met's Open Access Program. https://www.metmuseum.org/art/collection/search/255178
Holland, E., Jr., Dozier, L., & Holland, B. (1965). It's the Same Old Song lyrics © Sony/ATV Music Publishing LLC.
Klaver, I. J. (2018). River's paradox, river's promise: Meandering and river spheres. Open Rivers: Paradoxes of Water, No. 4. http://editions.lib.umn.edu/openrivers/article/
O'Connor, B. C., Kearns, J. L., & Anderson, R. L. (2008). *Doing things with information: Beyond indexing and abstracting.* Libraries Unlimited.
van de Passe, C. (1602–1607). Rijksmuseum. Amsterdam. Available through Rijksstudio program. https://www.rijksmuseum.nl/en/collection/RP-P-OB-15.946
Weiner, E. S. C., Simpson, J. A., & Oxford University Press. (2004) *The Oxford English dictionary.* Clarendon Press.
White, H. D. (2000). Introduction. In P. Wilson, & L. McCreery (Eds.), *Philosopher of information: An eclectic imprint on Berkeley's School of Librarianship, 1965–1991.* Bancroft Library.
Wilson, P. (1977) *Public knowledge, private ignorance: Toward a library and information policy.* Greenwood.
Wilson, P., & McCreery, L. (2000) *Philosopher of information: An eclectic imprint on Berkeley's School of Librarianship, 1965–1991.* Bancroft Library.

More Than Meets the Eye

2

2.1 Toward an Ontology of Proximity[1]

All of photography can be summarized as photons in, photons out. (O'Connor, 2009)

The very things which an artist would leave out, or render imperfectly, the photograph takes infinite care with, and so makes its illusions perfect. What is the picture of a drum without the marks on its head where the beating of the sticks has darkened the parchment? (Holmes, 1859)

Vision gives us proximity at a distance. Human vision is a matter of photons exciting receptors; so, in a general sense, photographs require little in the way of decoding. The photon data of any photograph maps the surface of what was in front of the camera with exquisite empiricism, much as does the human eye at any point in time (Fig. 2.1). This data shows us a great deal about those objects—color, texture, size, orientation—that enables interaction with our world. There is a difference, though, between the camera and the brain—the brain generally has a conceptual construct within which to make use of the eye's image. When we see the world around us, we generally do so across time—glancing at the thermometer, looking for the car keys, focusing on entering the road, shifting our gaze about the environment. Each scene is set within a larger context. Churchland refers to this as seeing "spatiotemporal particulars [within a] landscape or configuration of the abstract universals, the temporal invariants, and the enduring symmetries that structure the objective universe of [the brain's] experience." (Churchland, 2012) Any individual photograph presents an exquisite data set of "spatiotemporal particulars," but is, in and

---

[1] Adapted from Bonnici, L. & O'Connor, B. (2021). "More Than Meets The Eye: Proximity to Crises through Presidential Photographs." *Proceedings from the Document Academy,* 8:2, 2021. https://doi.org/10.35492/docam/8/2/14.

© The Author(s), under exclusive license to Springer Nature Switzerland AG 2022
L. J. Bonnici and B. C. O'Connor, *Proximity and Epidata,*
Synthesis Lectures on Information Concepts, Retrieval, and Services,
https://doi.org/10.1007/978-3-031-17094-2_2

Fig. 2.1 Photo mural *Meaning on the Wall* for Document Academy 2019

of itself largely bereft of universal particulars of either the maker or the seeker or the viewer. Any individual photograph has specific photon data, but any particular viewer may not have adequate contextual constructs to be able to make use (make sense) of the photograph. The photon data puts our eyes in proximity to the original scene.(Bonnici and O'Connor, 2019)

Photographs yield data based on physical proximity, but do not necessarily yield functionality for viewers with little or no contextual data. We use the term epidata for those synchronic attributes, those characteristics of a message that are subject to change with time and circumstance, and which are not readily available or easily knowable.

Probably many of us have engaged, unwittingly, in applying spatiotemporal particulars to abstract universals that translate to familiarities based on our brain's experiences. Have you ever gazed toward the sky on a partially cloudy day to see something familiar in the shapes of the clouds? We conjure up concrete items from abstract formations of water particulates rendered as clouds to the human eye (Fig. 2.2).

What do you see in this image? What do you make of this image? This is a snapshot of the sky rotated 90 degrees to the left. It was captured just a couple of hours after euthanizing a beloved pet. While walking a remaining pet, the dog kept looking up at the

Fig. 2.2 Cloud in the sky

sky. Following the pet's gaze, Laurie's eyes saw a heart-shaped cloud. Upon closer look, that same cloud looked a lot like her remaining dog walking as if leading her toward the sky. Was it the raw emotion of recently saying a final goodbye to a beloved pet of sixteen years that allowed the brain to interpret the cloud as something familiar; something endearing?

Without the epidata, what would you have seen in this image?

2.2   How Do Photographs Mean?

Words cannot describe photographs in the same sense that key words or subject headings can describe verbal documents because words are not native elements of photographs. It is quite likely that someone could say there is a lot of "red" in the photograph below; yet, there is no place in the image where the word "red" occurs. It is the case that there are many pixels with values (0n a scale of 0–255) close to Red 197, Green 19, Blue 36 that activate the human visual system in a way that is often labeled "red," but there are no letters "r," "e," "d." Words can describe epidata—reactions and associations that might be functional.

In 2004 we hung a 30-foot-long photomural in a gallery coffee shop, as pictured in Fig. 2.3. We then asked/allowed people to write on the print over a one-month period. We thought the physical presence of the collection as a whole together with social interaction might provide clues to what is needed/wanted for functionality of photographs. For DOCAM '19 we wanted to compare reactions to some of the same individual images in a mural format. In both cases we were interested in verbalization of reactions to photographs. Since words are not native elements of photographs but we often use words to describe our reactions to photographs, we looked to written comments as clues to reactions—threads connecting photon data to utility.

Beyond being able to see the photon data for a woman or a dog or a boat, how do we think about photographs? How do we look for an image that reminds us of summer time in Kansas, or makes us think of desolation or love? Might it help to know what the photographer was thinking? Would it help to know how others reacted? Does the size of a photo presentation matter? Does moving along a mural inspire a form of narrativity? We were curious to see if the necessity of moving (at least the eyes) along a set of images not intended to be a coherent narrative would still provoke some form of narrativity in the broadest sense. Would the bright red in both upper corners be noticed and commented on? Would the large number of hands get a mention? Would anyone ask why there are feet in the two bottom corners?

We had numerous comments written by a variety of viewers of the 2004 photomural and we wanted to see if we could find any commonalities between those and any comments from another photomural. The display space for the second piece precluded simply

Fig. 2.3 30-foot photomural with margins for viewer comments

hanging a new copy of the first mural. We selected a small subset of the original image set based largely upon the sorts of comments attached to them.

This in itself was intriguing; the process of mulling over the differences in the two installations was fruitful; and the process of preparing the reprise was at once delightful, thought provoking, maddening, and extraordinary. Engaging with the images again—both those from the first piece and the combination of new and re-used pieces in the second piece—was exhilarating and challenging. There is a good deal of personal history bound up in the individual images and in the making and showing of the first mural.

We prepared the mural for DOCAM '19 to use as a data gathering mechanism for our continuing work on proximity and epidata. Things did not go as planned. Unavoidable delays in hanging the piece and the venue (not the relaxed environment of a coffee shop,) seem to have dissuaded conferees from engaging with the mural with pens in hand. This was a disappointment. However, we soon realized that our planning of the new mural in and of itself provided a rich set of data and thoughts for sketching out an approach to photographs, epidata, and proximity.

2.3 Coding, Decoding, 'Spatiotemporal Particulars,' and 'Landscape[s] of Abstract Universals'

Understanding can be taken as the ability to do something with or because of a message— a means of bringing information to the point of use. We start with Shannon's cleaving of meaning from message. Some form of data is coded in some medium, transmitted, received, and decoded. Some forms of coding and circumstances of message making and decoding require little proximity of the recipient to the message maker, while some forms utterly depend on proximity.

Photon data from photographs presents to the eye/brain surfaces proximate to the camera at the time of coding the photograph; however, it does not necessarily present data on time, place, what was outside the frame, why the particular surfaces were selected, why a particular data recording technique was employed. In some circumstances, this may be an acceptable state of affairs. It may be possible that the simple surface data is all that is required by a particular user for a particular use. In other circumstances, data beyond the surface data may be utterly necessary. Epidata affords pathways, points to details that cast light on proximities that might otherwise go unknown. As we see in the tale of Theseus slaying the Minotaur, yet depending on Ariadne for his own escape from the labyrinth, even the most slender of threads can hold or bespeak significant proximity.

The proximity need not be to the original maker of the photographic message or even to the object recorded. It may even be that the original photographer will have a different decoding of the photographic message at some subsequent viewing time. An example of this is a photograph Brian's mother made in 1948. When Brian was about a year and a half old his mom made the picture in Fig. 2.4 of him with his dad on the steps of the art museum in Manchester, New Hampshire. A couple of years before his dad died, he went to sit on the steps just to think about the past. It occurred to him that he should make a self-portrait there by balancing his camera on a nearby fire hydrant employing the timer on the shutter release. When he showed his mom the new picture, she wondered if he could combine it with hers as a "way to think about the two times together, as if you were together 60 years apart." Figure 2.5 is the counterfactual image suggested by his mother. She was essentially proposing a form of temporospatial proximity. It was as if she were creating a memory that she knew would be important because of where she was in her own life.

Fig. 2.4 Brian and his father, 1948

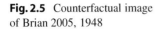

Fig. 2.5 Counterfactual image
of Brian 2005, 1948

2.4   Some Examples of Epidata and Thoughts on Possible Utility

The image in Fig. 2.6 appears in the lower left portion of our DOCAM 2019 mural. On
one level it is simply a photograph of somebody holding up blue reading glasses to a toy
animal's eyes. It might be useful at this level of decoding as a PowerPoint illustration for
"Pay close attention" or "Read your hunting regulations carefully." Those familiar with
DOCAM would likely recognize this as the antelope mascot of the group; while yet others
would know from having attended DOCAM '17 in Indiana, that this is Laurie Bonnici
holding up her glasses to the antelope and the photo was made by Brian O'Connor.
Knowing these details would give some context to the image of a smartphone making an
image of Michael Buckland, just to the right of the "FASH" photo in the center. Even
that sentence demonstrates the issues we are exploring because one would have to know
that Michael Buckland was in attendance at the Indiana gathering.

The "FASH" photo in Fig. 2.7 had appeared in the 2004 mural and was the object of
some speculation in written and verbal comments. In the earlier mural the image was not
quite so large and was not centered in the mural. The image is shown without cropping,
that is, the original did not present any additional surface data. Many viewers interpret the
straight-edged angle on top of the head to be a paper pirate hat or other such whimsical
head gear. "FASH" seemed perplexing to most who commented. The subject was actually
a much larger than life-size image of a model, cut out and pasted to a construction wall,
with weather and time having caused delamination around the edges—thus the pirate hat
look. FASH is simply the first part of FASHION COMPLEX. The image was made in
Las Vegas in 2003, when digital cameras were beginning to overtake film cameras. Brian
was attending a photo convention and was taken by the NIKON ads on the tops of taxis
"You have to love a town where image is everything!" The delaminating image seemed a
cute counterpoint to the ad. Having the freedom to shoot lots of images without worrying
about processing costs for film was part of the plan for taking the walk on which the

Fig. 2.6 Toy antelope and blue reading glasses

photo was shot. In retrospect, this entire paragraph is some metadata "…larger than life-size image of a model, cut out and posted to a construction wall…" with lots of epidata (nearly the remainder of the content).

At first glance the image in Fig. 2.8 might seem to be a simple, formalist photograph of a long hallway. As such it might be useful for representing the corridor of time, a long-term undertaking, a sad farewell, or just a pattern of straight lines broken up by a bit of human activity. For those who had their own physical proximity to the toy antelope above—folks who attended DOCAM'17 in Bloomington, Indiana, this might look familiar as the on-campus hotel, though not many folks keep track of hotel hallway design and décor. In the sense of presenting these images in the mural format, we wondered if there would have been a "narrative" connection made between this hallway photo and the antelope photo—would comments on the one informed understanding of the other.

Fig. 2.7 Delaminating billboard in Las Vegas

Would anyone have "mistakenly" assumed the interior image of the hallway and the exterior image in Fig. 2.9 of a city hall in New Hampshire were the same building because of their proximity in the mural? Likewise, would there have been any explicit connections made between hotel and city hall photos because they each have tiny human figures within largely architectural scenes? For that matter, would anyone comment—as did someone who happened to be near the city hall when the photo was made: "Looks like King Kong got the wrong building!" The distribution of similar colors and the converging parallel lines might set up a resonant pair for decorating a space. The exterior image is the city hall in Manchester, New Hampshire. Does that mean something?

This image in Fig. 2.10 has had many comments in other venues; one informal interaction at DOCAM '19 was similar to most of the other comments. On the surface, this is a white dog with pretty flowers, plant material, and assorted other objects. Terms such as pretty, beautiful, natural have been quite prevalent. When it is revealed that the dog has just been euthanized, most commenters have replied: "I am so sorry" or "How sad." Brian had the privilege of living with the dog for many years, so it is at once lovely and sad for him. The dog often watched over the yard where he builds kayaks, so there are pieces from one of the boats on her and the flowers come from her favorite parts of the yard. The image does not show that she is already in the bottom of a grave. Would that make a difference to meaning for some viewers? Would the context be a catalyst to build proximity if the viewer had also lost a dog to euthanasia?

Fig. 2.8 Hotel corridor in Indiana

Fig. 2.9 City hall in New
Hampshire

Fig. 2.10 Old dog at peace

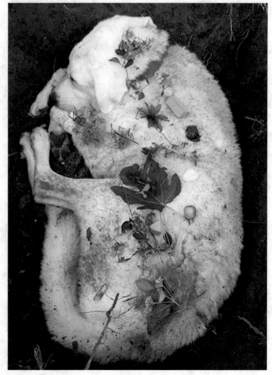

The grey scale image of two horses and a rider in Fig. 2.11 was in the 2004 mural, while the full color image was made 10 years later. They were placed together in our DOCAM mural to see if there would be comments on similarities. In both photos, the men are not competitors rather they are pickup men who remove some of the rigging from the bucking horse and help get the animal back to the chutes. This is a rather mundane part of the rodeo for many viewers, yet a task requiring skill, strength, and agility.

The top image was used in previous work (O'Connor et al., 1999) in which we asked viewers to describe images and then asked cataloguers to describe the same images. Viewers tended to describe actions and apply adjectives—"rugged," "strong," "amazing skill"—while cataloguers tended to apply nouns—"Cowboys," Horsemen and Horse-women." The postures of the riders are remarkably similar even though the one in the top images is working on a horse whose head is toward the camera, while the one on the bottom works on a horse whose tail is toward the camera. To whom would the difference

Fig. 2.11 Two rodeo scenes

matter? Does the color add or detract for some particular use of such an image? For whom would it matter that one is in California and one is in Texas?

The photograph (Hillard, 1864) in Fig. 2.12 is quite ordinary on its surface, perhaps even clumsy looking with the inclusion of someone's fingers. Anyone can do an online search for Samuel Downing and find a "better" version of this photograph. This photograph was made explicitly to represent the materiality of this particular version of the photo of Samuel Downing. It appears in an original edition of *The Last Men of the Revolution. A Photograph of Each from Life.* (Hillard, 1864) published in 1864. Rev. E. R. Hillard interviewed the last seven living veterans of the American Revolution and made a photograph of each one. In 1864 there was no half-tone printing or any other method of rapid printing of photos in a print run of a book. Each of the photographs in all of the copies is handmade and glued into place. So, when you are touching the photograph you can feel that it is a separate piece of material; also, you are touching a print made by someone who chatted with veterans of the Revolution and some of the last people to have lived in a time before photography.

Fig. 2.12 Hand processed print of Samuel Dowling, veteran of the American Revolution

2.5   Photographic Proximity

The comments made about the photographs from our 2019 mural presented at DOCAM illustrate that having a picture's photon data strike the eye/brain of a viewer is not enough; exquisitely specific data without some sort of context for the data, some form of proximity beyond the utterly specific photon data may lead to limited functionality of the photographic document.

We suggest that a first order taxonomy of proximity comes into play. For some uses, very general notions of what surfaces and what coding practices recorded a particular set of surface data: cute photo of reading glasses on a toy antelope; for some uses, a deeper level of specificity: DOCAM '17 in Indiana; and for other uses, even more specificity, of a sort that might be hard to come upon: glasses and antelope photo was made by the authors of this piece, the glasses had made their first appearance at DOCAM '16 in Denton, Texas, and disappeared a little later, only to reappear a couple of years later.

We see this play out in many photos posted to social media platforms. Albannai presents several examples of conversational use of photographs on Facebook (Albannai, 2016) that speak both to threads of proximity and to the collaborative environment in which we can now discover, request, and contribute to such threads. This photograph in Fig. 2.13 (Mydans, 1936) from the Library of Congress (labeled "Cigar store Indian, Manchester, New Hampshire") was part of a discussion thread on a Facebook page concerned with the history of the city. Comments included: "My mom said this was on the corner of bridge and elm st," "She is 95 … she remembers a man named Bernie who worked there;" "my Dad used to stop there for his newspaper, of course smokes, and always treat for me and my sister;" and "My mother took me there to buy cigars for my father for Christmas."

We have here locative information that does not exist in the Library of Congress record, and we have threads of how the store fit into the life of the city and how it was significant to the city's inhabitants. Perhaps most demonstrative of the notion of threads of proximity is this comment on the same picture: "Saw that photo on a wall in Denny's [restaurant] in CA a couple years ago. Asked our young waiter if he knew where it was taken and why. He didn't so we told him. He found the story interesting and promised to share it with the rest of the staff"—note Denny's in CA is about 3000 miles from the corner of Bridge and Elm Streets, in Manchester.

The examples we presented and the work of Albannai demonstrate three levels of generality: any image with a particular object or characteristic will do; an image with certain qualities and intriguing connections is wanted; the photograph must show particular qualities or have a backstory that is explicit as to why this is the most useful image. Yet each those three regions on a spectrum of utility can function along several vectors. We might here expand the idea of a taxonomy of proximity into the more inclusive, environmental notion of an ontology of proximities—referring to types of connections, types of uses, and circumstances of discovering the threads.

Fig. 2.13 Carl Mydans photo "Cigar store Indian, Manchester, New Hampshire" for U.S. Resettlement Administration, 1936

2.6   Proximity to Crises Through Presidential Photographs[2]

We look at three photographs, each made at a time of profound crisis, in order to tease out notions of proximity. Each image was made by a highly skilled photographer, but each presents the photon data from only a fraction of a second. How is a viewer to insert the spatiotemporal particulars of that faction of a second into their own abstract universals? Can words and other images from the photographers enhance the viewer's proximity to the original? Can we make use of the photographers' accounts of their proximities for enhancing the understanding of individual viewers?

---

[2] Adapted from Bonnici, L. & O'Connor, B. (2021). "More Than Meets The Eye: Proximity to Crises through Presidential Photographs." *Proceedings from the Document Academy,* 8:2, 2021. https://doi.org/10.35492/docam/8/2/14.

2.6.1   Crisis—1963

In 1963, light generated photon data from a group of people in a small space. We can see some of that photon data today in the photograph in Fig. 2.14. It was made in the midst of an extraordinary crisis in the United States. The photo was made not only for news value and history but also to resolve one aspect of the crisis—continuity of government leadership. Yet, it shows essentially nothing of the actual crises. Photon data brings a viewer into a form of close proximity with a portion of the original data of the scene. However, without metadata and without epidata (contextual information not ordinarily collected), the functionality of that proximity may be severely limited, obstructing access, use, and understanding. The photograph, in and of itself does not tell us what crisis is documented; nor does it tell us just whom we are seeing or what is happening. This becomes increasingly problematic as spatial and temporal distance from the event increases; thus, requiring means for facilitating proximity become more significant for understanding.

Fig. 2.14 Swearing   in   of   Lyndon   B.   Johnson   as   President,   photograph   by   Cecil Stoughton (Stoughton, 1963)

Metadata, such as captions, can add functionality and enhance the likelihood of under-standing. A caption presents words that give some context: "Cecil Stoughton's photograph of Lyndon Johnson being sworn in as President of the United States aboard Air Force One immediately after the assassination of President Kennedy." Some metadata may be less than functional in terms of the relationship of such a photograph to the crisis from which it emerged. The Library of Congress (LOC) Subject Headings applied by the Prints and Photographs division give no hint of the assassination of Kennedy being the primary circumstance of Johnson's inauguration:

- Johnson, Lyndon B.–(Lyndon Baines),–1908–1973–Inaugurations
- Onassis, Jacqueline Kennedy,–1929–1994–Public appearances
- Presidential inaugurations–Texas–Dallas–1960–1970
- Oaths–Texas–Dallas–1960–1970.

The folder holding this photograph in the John F. Kennedy Library is titled in an almost bizarrely comical way: Trip to Texas: Swearing-in ceremony aboard Air Force One, Lyndon B. Johnson (LBJ) as President.

The obituary for the very first official White House photographer Cecil Stoughton in the *New York Times* enhances proximity to the image and its impact. Within a recounting of Stoughton's life, with considerable attention to his role as the first official White House photographer, two sentences tell the reader just why the picture was so important:

Mr. Stoughton's picture is the only photographic record of the Johnson administration's abrupt, official beginning. At a precarious moment in the country's history, it gave the pub-lic at least a semblance of continuity: one president sworn in as the widow of another looked numbly on. (Fox, 2008)

What we do not see in captions or in the Library of Congress Subject Headings (among others) or even in the *New York Times* obituary, are: the efforts of the photographer to be on the scene; the necessity to reassure the American public; why the image is black and white; the purpose behind the composition. Not knowing such contextual information decreases the likelihood of finding and understanding a photograph, even one so inti-mately connected to a horrific and consequential crisis. We look to the backstory on the production of the image to glean epidata—the not so easily known or immediately avail-able clues to understanding. These illuminate the proximity of the photographer to the subject/event, thus contextualizing the visual proximity the photograph presents. Epidata may not be useful to all potential users, but they may well be crucial to some viewer's understanding.

We ask: What are the things that are knowable about a photograph beyond title/caption/photographer and metadata such as time, place, camera? How can we under-stand photographer decisions, mechanical constraints, cultural constraints? What can be gained by some form of proximity to the photographer's initial making of an image and by some form of proximity to the circumstances of the image being published? While these may seem primarily of interest to photographers, they may well clarify matters for some viewers. Why do we not see the face of the federal judge administering the oath—the first female judge to do so? Because there was no angle in the small room from which to make an image of Johnson and Kennedy and also include the judge's face. Knowing that time was of the essence and it was only possible to make 21 pictures, might explain to a historian why certain people are in the picture but not well lighted. Knowing that there was no professional audio recorder available but somebody remembered there was a dictating machine in the Air Force One office explains the little square item held in front of the judge. A fashion historian might find the clothing of political figures of 1963 of interest, and perhaps especially Jackie Kennedy's jacket. Stoughton's image of the jacket is both a comment on his skill at choreographing the subjects and a cautionary tale about accepting a single photograph as the record of the moment. This photograph had to reassure the nation, so Stoughton had her stand with her left shoulder to the camera—the side farthest from the president when he was shot, the side with the least amount of bloodstain. Lady Bird Johnson's diary entry for that day notes the reality of the other side of the jacket and the skirt:

> Her hair [was] falling in her face but [she was] very composed ... I looked at her. Mrs. Kennedy's dress was stained with blood. One leg was almost entirely covered with it and her right glove was caked, it was caked with blood – her husband's blood. Somehow that was one of the most poignant sights – that immaculate woman, exquisitely dressed, and caked in blood. (Johnson, 2001 - as quoted in *Lady Bird*)

Since the dress is held by the National Archives—uncleaned—and, by the wishes of family, secured from public view at least until 2103, photographs are the primary source of study of the outfit for any purpose.

We propose mechanisms of proximity, constraints on proximity, and levels of proximity in the making of the photographs as a substructure for connecting a viewer's abstract universals with those of the photographer. We assert that the functional strength of the thread of proximity depends on the partners at both ends of the thread—the coding practices and the decoding abilities.

We should note that every president since Kennedy has had an official White House photographer. Before Kennedy, White House photographs were made by members of the Army Signal Corp essentially as photographic records of events such as hosting dignitaries, signing legislation, and the like. Since Stoughton's time the official photographer has had onsite office and facilities, but each president has used the photographer's talents differently. There is not a statement purpose of the photographer's position or, indeed, of responsibilities that can be assumed to hold across all the presidential photographers. Some have had close relations with the president and photographed casual and intimate moments, where others have essentially been record makers (Krule, 2012). The photographs made by the White House photographer are publicly available with certain constraints on classified materials, but how many photographs are made of what sorts of topics and under what constraints vary with the president. The photographers are proximity pieces/agents between the president and the public, but without knowing the nature of the photographer's link to the president, we cannot know the nature of a viewer's proximal link to presidential events.

2.6.2 Mechanisms of Proximity

Representation

We propose that photographs and perhaps documents in general are mechanisms that resolve the past, in the sense of (re)presenting its constituent parts. We can ask about any document what sort of resolving power does it afford one in determining a past state? A photographic document recovers a vector state of the past that enables a close mapping of surface qualities. There exists the possibility of recovering from the initial files, the temporal, spatial, and spectral component(s) of some State 1 from the State 2 represented in the photograph.

Cecil Stoughton was witness to a crisis and provided some degree of eyewitness presence for future viewers. Precipitating event for the crisis was the assassination of President Kennedy—Stoughton's photographs show us nothing of that event. The crisis at hand was assurance of continuity of government—Stoughton's photographs record the swearing in of the new president. We should note that when we presented at DOCAM '21 and asked if anyone was familiar with the photo, the comments mentioned Kennedy assassination; yet the image does not directly document that event. It documents outcomes/ensuing activities of the assassination.

2.6.3 Photographic Processes at the Time of the Crisis

In a time when nearly everybody has a camera in their cell phone that is capable of making color photographs and videos, it may be difficult for today's viewers to comprehend that in 1963 there was only one camera aboard Air Force One. Cecil Stoughton—the official White House photographer—was photographing with a Hasselblad 500C camera, similar to the camera in Fig. 2.15, that weighed almost four pounds and could make only twenty images on each roll of film.

To provide proximal links for current viewers of this photograph, we turn to Stoughton's words on the making of the photograph under emotional stress, the time stress of knowing engines on Air Force One were already revving to leave for Washington, DC, and having someone on a phone in Washington, DC dictate the words of the oath of office.

Fig. 2.15 Hasselblad 500C
camera, photograph by
Jonathan Mauer

I took the color film out of my Hasselblad and reloaded it with black and white, [since it took] *two hours to process color film in those days, they don't transmit color on the wire photos, and black and white's the only way to go.*

And now I'm in the cabin where the oath's going to be taken, and the president says, "Cecil, where do you want us?" you know. Because I had to arrange to make sure that I'd get the necessary picture: him holding his hand up and his other hand on the Bible, and anybody that would be surrounding him would be important, like Mrs. Johnson and Mrs. Kennedy.

The judge read the oath, and the president repeated it. I was clicking pictures left and right, standing in my little leather seat and spraying around the cabin while they were doing the talking. Got the picture that was required: Hand up, hand on the Bible, eyes open. And Jackie and Mrs. Johnson on the other side. And then the president said, "Let's get this plane back to Washington."

Well, I couldn't go back to Washington with it, because I had to take the film off and get the film processed and put it on the wires for the wire services, because the world was waiting to see what was going on … So it was important that the film get processed and released as quickly as possible.

…, went to the AP photo lab downtown in the Dallas Morning Herald's photo lab, I think it was. We processed the film, and made the prints, and put it on their wire service, wire photo drums. I repeat this drums thing, and a lot of times it's not knowledgeable to people who weren't aware of what the drums were. But it's a telephonic device for transmitting pictures through the telephone wire.

So the picture was developed and processed, printed and transmitted, and reprinted in Washington. By the time Johnson was getting off the plane at Andrews Field two hours later, the picture was on his TV screen in the cabin of the plane, and he was watching himself being sworn in before he got off. (Stoughton, 2002)

2.6.4   Being There: Proximity to the Kennedy Family

Preparation

Captain Cecil Stoughton, seen in uniform in Fig. 2.16, was trained as a US Army photographer and filmmaker under some of the luminaries of the day, such as Alfred Eisenstadt, Margaret Bourke White, and Ronald Reagan. He had seen combat action, including photographing events at Guadalcanal. He had considerable technical expertise and the ability to work under pressure. Assigned to photograph the inauguration of John F. Kennedy, he produced images that caught the eyes of the new President and First Lady. Kennedy arranged for him to be assigned as the first official White House photographer, with an office in the White House and a dedicated telephone in his home.

Fig. 2.16 Photographer Cecil Stoughton and wife, Faith Hambrook Stoughton, Attend Military Reception at the White House

In his time as White House photographer, Stoughton:

sat poised each day for the sound of a buzzer, which meant President John F. Kennedy was
ready for his services. Over 35 months, Mr. Stoughton shot state dinners, receiving lines and
visitors of all kinds, from foreign leaders to "50 singing Nuns," …But when the visitors left,
Mr. Stoughton had the chance to capture the First Family in far more personal settings – in
their White House quarters, at their vacation homes and on their many travels. (Fox, 2008)

2.6.5   On the day of the Assassination

Stoughton was with the press corps in the presidential motorcade and once the sound of
a rifle shot was heard, he made use of several proximities to arrive at Air Force One as
the only photographer.

Stoughton's words on making his way:

The driver of our car was a local police officer. … He recognized somebody on the sidewalk
there; he said, "What happened?" And he said it sounded like–looked like somebody got shot
in the president's car, and they must have gone to Parkland [Hospital.]

  I told the guy, well, "Let's go. We need to get there, too." So we took off real quickly. …
ended up at this Parkland Hospital. Jumped out of the car and started making pictures … The
president's car was in a little emergency ambulance drive-in. I went on inside the hospital–
being part of the staff, I was not precluded from going in …

  out of the comer of my eye I could see Johnson, Vice President Johnson, and Lady Bird
and Rufus Youngblood, his Secret Service guy, walking rather rapidly towards the door that I
had come in just a few minutes before. And this chief warrant officer that was handing me this
phone, I said, "Where's he going?" And I nodded my head like that. He said, "The president's
going to Washington." …and my realizing immediately that Kennedy wasn't the president
anymore and that Johnson was, nominally, and knowing that there was a need for a ceremony
of some kind, either impromptu or official, it behooved me to be with him. So, when he said
the president's going to Washington, I said, "So am I."

  I didn't ride out in Johnson's car, but there was another car, police car, there, staff car, so
to speak. – got into this car with a driver and followed the Johnson party out to what turned
out to be Air Force One. Kilduff, Malcolm Kilduff, came running up the aisle and said, to the
effect, "Thank God you're here, Cecil. The president's going to take his Oath of Office on the
plane. You're going to have to make the pictures and release it to the press because (a) there's
no room and (b) they're not here anyway." (Stoughton, 2002)

Physical proximity became an issue after the prints had been sent over the wire because
of the blood on Jacqueline Kennedy's outfit. Kenny O'Donnell at the White House had
evidently seen other images showing the blood-stained clothes; so, he sent a plane to get
Stoughton back to Washington, DC, where Stoughton went to his darkroom and made
prints of the images that had been released—showing that there was no blood because

he had posed her and framed the image to avoid the blood. In 1963, printing from the negative required considerable time and required the negative, the piece of film that had come from the camera. To show the White House staff that his negatives showed no discernible blood, Stoughton had to be in Washington, DC.

In the case of Stoughton's photograph of Lyndon Johnson taking the oath of office, there are several other photographs to expand the spatiotemporal particulars, to expand a viewer's proximity to the event. A sample of the other images on Stoughton's roll of 21 pictures publicly available through the National Archives and Records Administration is shown in Fig. 2.17.

Putting Stoughton's published image (lower right) into proximity with a sample of others made seconds before and after gives a sense of his observational acumen at the time of making the image and in preparing it for publication. All the people in the photographs are in almost exactly the same positions, and the camera is nearly locked into the same position because of the large crowd in the small space. Yet, Stoughton was able to record

Fig. 2.17 Sample of other images on Stoughton's roll of film from the swearing in of Lyndon Johnson

faces of most participants, even those in the background. He was also able to frame Lyndon Johnson in the center to give him a look of authority, while still framing the judge and Jacqueline Kennedy prominently.

The other two photographs give a sense of just how crowded and chaotic the situation was. In his comments, Stoughton mentions "spraying around the cabin"—photographing everyone that was on the scene; we actually see some different people in the different images.

Having these is not likely to make the most prominent image more readily accessible in a search; however, they do present the crowded confusion during the few minutes surrounding the oath taking photograph, perhaps increasing the utility of the prominent image. It might be said that the primary image could be an accidental, inadvertent link to the behind-the-scenes images and to just who was in the space.

Next, we turn our lens of proximity and epidata on a more recent event, yet one that documents history through the lens of a White House photographer. We see an image of another president in another crowded space with other people. But rather than being the subject of the unfolding event, the image documents the people viewing the event as it was unfolding.

2.6.6   Crisis 2011

In 2011, Obama's White House photographer Pete Souza—who had also been a White House photographer for President Reagan, gathered the photon data of several people gathered staring in a small room, in Fig. 2.18. Souza was using a digital single lens reflex camera with a wide-angle lens—a Canon D Mark II, similar to the camera in Fig. 2.19, though with a different lens—that weighed about the same as Stoughton's Hasselblad, yet capable of making hundreds of high-resolution full color images on a single storage card. Souza had been a newspaper photographer in Chicago and had covered Obama's career there; he had also photographed events in Afghanistan immediately after the events of September 11, 2001.

As with the Stoughton photograph, in and of itself, Souza's image does not tell us what crisis is represented; nor does it tell us just whom we are seeing or what is happening. Souza's photograph, made nearly 50 years after Stoughton's photograph of President Johnson, affords rich comparative analysis. In many ways it is almost the same picture. We do not see anything of the crisis involved. The photo was made from the same corner of a small room, with lots of people, the angle of the wall trim in both pix is the same, and Jackie Kennedy and Hillary Clinton occupy both the same space and hold the role of 'punctum'—not the primary object but one that gives the emotional punch. (Johnson, 2011) Even what is different is similar—the space containing the judge and LBJ is totally

Fig. 2.18 Pete Souza photograph of the Situation Room during raid on Osama Bin Ladin, 2011 (Souza, 2011)

Fig. 2.19 Canon EOS 5D Mark II

empty in the Obama photo, simply inverting the primary subject of LBJ to negative space. Of course, much of this is simply due to the fact that many impromptu presidential events take place in small places and there are often many folks involved. Perhaps the most obvious difference is that Souza's photograph is in color.

Context provided by epidata shows significant similarities and significant differences in the circumstances of production, in the role of the photographer, and in the initial intention/use of the photograph. There are two primary sources for the epidata—both of them substantially different from Stoughton's oral history. Souza maintains an Instagram account with more than 2,000,000 followers. On that account he has posted a 22-min video in which he gives the background of making the "Situation Room" photograph. He also maintains a site on Flickr.com with 6668 photographs from the Obama White House. Flickr has an EXIF button, as shown in Fig. 2.20, that enables display of a photograph's Exchangeable Image File format data. Ordinarily, EXIF data is used to store technical information generated by the camera at the time of exposure, e.g.: time, GPS coordinates, length of exposure, camera model, lens type. EXIF data can also hold notes entered by the photographer.

Souza's EXIF note is 372 words; many of the words simply tell us who is pictured in the photograph, but there is a good deal of contextual information also:

May 1, 2011 "Much has been made of this photograph that shows the President and Vice President and the national security team monitoring in real time the mission against Osama bin Laden. Some more background on the photograph: The White House Situation Room is actually comprised of several different conference rooms. The majority of the time, the President convenes meetings in the large conference room with assigned seats. But to monitor this mission, the group moved into the much smaller conference room. The President chose to sit next to Brigadier General Marshall B. "Brad" Webb, Assistant Commanding General of Joint Special Operations Command, who was point man for the communications taking place. With[sic] so few chairs, others just stood at the back of the room. I was jammed into a corner of the room with no room to move. During the mission itself, I made approximately 100

Fig. 2.20 EXIF tab on Pete
Souza's Flickr page

Canon EOS 5D Mark II

EF35mm f/1.4L USM

f/3.5   35.0 mm

1/100   ISO 1600

Flash (off, did not fire)   (i) Show EXIF

photographs, almost all from this cramped spot in the corner. There were several other meetings throughout the day, and we've put together a composite of several photographs (see next photo in this set) to give people a better sense of what the day was like. [Names] Please note: a classified document seen in front of Sec. Clinton has been obscured."

At the beginning of the note, we are given more situational detail than the Library of Congress Subject Headings provide for the Stoughton photograph of Johnson, together with an acknowledgement of the public reception of the photograph. We learn the White House Situation Room is more than one small space and why the photograph was made there. Unlike Stoughton, who choreographed the Air Force One image, Souza was "jammed into a corner of the room." He mentions that he made "approximately 100 photographs" of that particular meeting and made images of other meetings from which he and his staff "put together a composite …to give people a better sense of what the day was like."

The image in Fig. 2.21 shows a frame from a video on Souza's Instagram account. We learn from that video more details about the circumstances of this particular image. Unlike Stoughton's situation, Souza had advanced notice that "something historic" would probably be happening "Saturday or Sunday." He was not told just what it would be, only that he should be available. On the two days before the weekend, Souza accompanied Obama to Tuscaloosa, Alabama to observe severe weather damage, Cape Canaveral for a rocket launch that was delayed by weather, and a college commencement where the President gave the address—all the while knowing "something historic" was to happen on the weekend, but not yet knowing what.

When the event commenced, Souza walked with Obama to the Situation Room, chatting about the matter. In a preliminary meeting, he made 140 images, during the event he made about 100 images, then in subsequent meetings on how to break the news and whom to contact he made several hundred more photographs—for a day's total of 1003. We learn that he only made about 100 images because his camera was quiet but not entirely silent and he did not want to disrupt the concentration in the room. He was shooting with a digital camera capable of making some hundreds of images on a single storage card, yet for the time he was in that room he was making photographs at the same rate to which Stoughton had been constrained. Again, while the moods in both situations were serious, the Situation Room mood required the least possible photographic interference.

While Stoughton rushed from Air Force one to a Dallas newspaper darkroom in order to get his photo "on the wire" as quickly as possible, we learn from the Instagram video that Souza gave his files to the White House photography staff and went home sometime after midnight. He returned early in the morning to do a "rough edit" of the 1003 down to 50 images that best gave a sense of the event. He printed the one now known simply as the Situation Room photograph and noticed what seemed as if it might be classified material in front of Hillary Clinton. This required confirmation—it was classified; it required consideration of whether the picture could be released to the public with the classified portion digitally obscured—a possible violation of the concept of not altering

Fig. 2.21 Pete Souza on his Instagram page (Souza, 2020)

White House photographs. As we see in the EXIF data, it was released with the alteration; in the Instagram video we learn that it was an extensive discussion and the only photograph ever released that way from the Obama White House.

Also, in the Instagram video Souza explains that he is often asked exactly what moment in the raid on bin Laden is recorded in the Situation Room photograph and that he cannot say. He has the time stamp on the image file, but until the official timeline of the operation is declassified, he cannot link the operational time to his file's timestamp.

2.6.7   Spatiotemporal Particulars

Ken Johnson wrote in 2011: "Rarely has a photo revealed so little while evoking so much. It shows an intent President Obama and other officials in the White House Situation Room, but tells little about what exactly the situation is, except that they are watching something off to the left." (Johnson, 2011) This leads us back to questions of what are we seeing? What do we need to know to know what we are seeing? What did the photographer mean for the intended audience to see?

Immediately after the Kennedy assassination, the White House needed a photograph to assure the world that, despite the shocking death of the president, there was continuity of government; immediately after the bin Laden raid, President Obama called several world leaders and made a televised speech—there was no pressing need for a photograph of the Situation room to calm fears. It should probably be noted that some quarters have asked for a photograph of bin Laden's body, but that is another issue.

When Ken Johnson was writing his article about the photograph, he could assume most who saw the photograph would recognize Obama, but that assumption might not hold so strongly as time passes. As the event pictured recedes into the past, fewer viewers are likely to have the situational particulars to associate the pixels of the image with particular details of experience. Think of the days when photo prints were put into albums—a set of particulars themselves receding into history—how many people experienced the frustrations of trying to remember "Was that my 18th birthday or 19th?" or "I wish someone had written down who the woman on the left is." Recently we were showing the Stoughton photographs to some people in their 20s, many from countries other than the United States; when we asked "What does this mean?" one student responded: "That is not in our history books." A lovely demonstration of the need for context beyond names of objects in the photographs.

We can look at the additional situational particulars provided by the photographer to explain more to those of us who were witness, in any sense, to an event and to provide proximal bridges for those who were not witnesses. For those of us who were high school students at the time of the Kennedy assassination, the Johnson image is likely still a significant spatiotemporal particular—a direct link to the memories of that day; for those somewhat younger, the picture may be a "my parents told me they could remember exactly where they were when they heard the news" particular; for younger viewers it may be a link to a history lesson or trip to Dallas or even simply "some people in a room."

2.6.8   Proximity, Epidata, Spatiotemporal Particulars

We can see in the EXIF data for Souza's "Situation Room" that the shutter was open for 1/100th of a second—a very short time at human scale. We can ask: Is that enough time? To which we might well respond: "Enough time for what?" That moment's worth of photon data shows who was in the room, though it tells neither names nor titles nor functions. Serious concentration is evident, though the picture alone does not tell us what sort of event is requiring the serious attention; nor does it tell us how long the attention has been so rapt nor whether the previous or next moment would have shown excitement.

One might ask why not make video and audio recordings? At the time of the Johnson inauguration photograph, the preparation for making a 16 mm film with audio would likely have taken longer than the entire time the plane sat ready to leave; it would have required a large camera on a tripod and bright lights—which the limited space on Air Force One could not accommodate. In the case of the Obama photograph, a reasonably high-quality video could have been made with essentially the same equipment as made the still photographs, but narration of the event and accompanying discussion of strategies would likely have to be classified, possibly rendering the video unsuitable for release to the public.

Another issue arises out of the technology—in the Stoughton photograph the viewer has no color data. For the original purposes of the photograph this is of little consequence; however, even at the time, Jackie Kennedy's fashion was a matter of public interest. Especially after her suit was covered in her husband's blood, the color became a matter of intense interest and even some controversy over whether or not any blood shows in Stoughton's photograph.

We tried an experiment and ran Stoughton's photograph through three online colorizing engines—each used artificial intelligence and none required more than 10 s to produce results. As can be seen in a glance at Fig. 2.22, results varied in quality of coloring within the lines—recognizing the boundaries of discreet objects—but they all looked like color photographs. However, not one rendered Jackie's coat as raspberry pink.

We do have a Cecil Stoughton color photograph of the raspberry ensemble at the time of arrival in Dallas, shown in Fig. 2.23. We could look at lack of color data and at the lack of the motion and audio components of events as technological distance or weakening of the possible proximal ties between a record and an event. Other photographs, personal accounts, and even simple understanding of the recording processes can alert us to what is lacking—sometimes filling in what is missing, sometimes only acting as a caution to interpretation.

Even these simple colorized images raise a growing concern in trusting photographs to present photon data of a significant event—deep fake proximity. Deleting people and objects from photographs or adding them in has been practiced since the mid-nineteenth century. Analog techniques could be convincing to the casual glance, but were generally detectable with some close scrutiny. Digital additions, deletions, and alterations can be nearly undetectable and are the subject of considerable research. Digital forensics research

Fig. 2.22 Three AI colorized versions of Stoughton photograph of swearing in of Lyndon Johnson

Fig. 2.23 President John F. Kennedy and First Lady Jacqueline Kennedy Arrive in Dallas, photograph by Cecil Stoughton

Hany Farid notes: "Stalin, Mao, Hitler, Mussolini, and other dictators routinely doctored photographs so that the images aligned with their messages…They knew if they changed the visual record, they could change history." (Farid, 2019) Digital manipulation requires "…techniques for reverse image searches, metadata analysis, finding image imperfections introduced by JPEG compression, image cloning, tracing pixel patterns, and detecting images that are computer-generated." (Farid, 2019)

Through close analysis of two famous presidential photographs, we have demonstrated the value of proximity—the proximity of a recording device to the event, the representational abilities and constraints governing the richness and functionality of the representation, and the contextualizing roles of various sorts of epidata—to understanding and instilling relevancy over time.

We emphasize that both of these photographs were made as the only records and each by the only individual with a camera at the event. Now, 58 years after the Johnson swearing in photograph and 10 years after the "Situation Room" image, it is routine that many more recording devices are typically on scene. The *New York Times* produced a visual investigation of the events of January 6, 2021 in Washington, DC, of which it said: "A six-month Times investigation has synchronized and mapped out thousands of videos and police radio communications from the Jan. 6 Capitol riot, providing the most complete picture to date of what happened — and why." (Khavin, 2021)

Most of those recording devices are sophisticated and highly portable and capable of recording video, sound, and GPS location data. This promises more sorts of data, triangulation of representations, path tracking, and multiple representational agendas—news media reportage, participant eyewitness views, police recordings of events and reactions. Intriguing potential is balanced by two points: each of the recording devices has its own limitations; the agglomeration of all the representations requires organizing principles and, perhaps, selection practices that impose their own distancing from individual perceptions of the events. Ensuing archival processes are impacted by these limitations, as archival practices themselves can pose limitations depending on purpose, space, and available metadata.

We assert that the greater the distance—temporal, spatial, cultural—a person holds from an event, the greater the need for connecting threads between that person's abstract universals and the particulars of the event. A photograph is an exquisite representation of the surfaces in front of the camera; yet, as with all representations, the highlighting of certain attributes, necessarily means leaving some behind (Marr, 1982). Some of the attributes left behind are due to technical capabilities of the recorder and some are the choice of the photographer. For a photograph to function as one of those connecting threads a viewer must know both the representational capabilities of the medium at the time of making, together with the intended purpose. As with any representation, the functionality of a photograph depends on the viewer understanding the making of the representation.

Another set of photos, of a president in a time of crisis were made under very different circumstances. While similar representational issues hold for these, the very different circumstances of the image production afford a set of spatiotemporal particulars that form, in a small way, an abstract universal.

2.6.9   Crisis 2001

An intriguing companion for the Stoughton and Souza photographs is the set of images made of President George W. Bush in a Florida classroom for a routine visit to promote reading. As seen in Fig. 2.24, there are many students, advisors, and media professionals. There is nothing about the photograph that "says" George W. Bush. We can see words on objects in the photograph—White House—though they are not native elements of the photograph. This single photograph tells us nothing about a crisis developing. The significant, intriguing difference from the Johnson swearing in ceremony on Air Force One and Obama in the Situation Room is this event was recorded by multiple cameras—both still and moving images.

If one does a Google search on "George Bush in classroom 9/11" several images come up, as in Fig. 2.25. A screen shot of some of those images shows different moments—Chief of Staff Andrew Card's head at different distance from Bush's head—and different angles and different framing. Audio was recorded during the event, but what Card whispers is not audible. We know now that this is the moment President Bush receives word of the terrorist attack on the Twin Towers in New York; yet there is still nothing in the official White House photographs or the media images that "says" terrorists or Twin Towers. For all the extraordinary specificity of the images with their recording of the photon

Fig. 2.24 President George W. Bush in Florida classroom on September 11, 2001

data of the moment, captions and the stories within which the images are imbedded are necessary for linking the spatiotemporal particulars to the abstract universals of "what's going on here?".

A few minutes after President Bush had finished his reading with the school children in Florida, he gathered with advisors in a nearby classroom to get the latest intelligence on the events in New York. One of the White House photos of that time—visible in the top row of images in Fig. 2.25, second from the right—shows another portrait of a president at work during a crisis. This photograph differs from others we have discussed by actually presenting a small glimpse of the crisis at hand. In Fig. 2.26 we see a figure pointing behind President Bush toward a television screen showing an image of the Twin Towers in New York with a plume of smoke billowing out. The screen occupies only a small fraction of the area of the image and the screen is partly obscured by light reflections; yet, an image of the crisis itself—or, perhaps more properly, an image of an image of a fraction of a second within the crisis—is there.

Twenty years later, we find through online search tidbits of proximity data scattered throughout search results. Data that offers more than the eye can see in the images. Much of this data has been collected over the ensuing years, offering proximity to the viewer. Much like the Johnson swearing in event, this proximity data serves to remind and inform people across generations. In sum, epidata serves to keep the photon data functional.

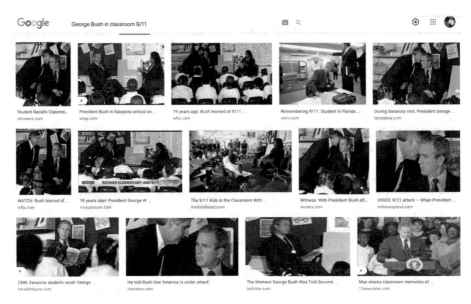

Fig. 2.25 Sample of image results for Google search on "George Bush in classroom 9/11"

Fig. 2.26 President Bush and advisors receiving intelligence, 9/11, 2001

2.7   Personal Presidential Photograph Epidata

In 2012 the *Ms. Magazine Blog* posted a piece on Barry Goldwater as a conservative Republican who had supported women's reproductive rights. (Chesler, 2012) Illustrating the post was the photograph in Fig. 2.27 of Goldwater during the 1964 presidential campaign, Lyndon Johnson's first election since the assassination of John F. Kennedy.

One might wonder just why this picture of Goldwater was chosen to illustrate the article. The image clearly has crease marks and was made outdoors. The image in Fig. 2.28 by Marion Trikosko for *U.S. News & World* Report shows Goldwater a few months later, shortly after his losing the presidential election. It is a more standard professional image available on Wikimedia with no known copyright restraints. It might be that in 2012 the author simply knew Flickr.com was a place to go to find images with Creative Commons sharing agreements.

The photo credit at the bottom of the article reads: Photo of Barry Goldwater from Flickr user *zudark* under *Creative Commons 3.0*. If one goes to Flickr and seeks out the image from "zudark" one sees something rather more complex. Zudark's page on Flickr.com actually notes that the photograph was made by "pelicanwind," aka Brian O'Connor.

Fig. 2.27 Barry Goldwater photograph by Brian O'Connor in *Ms. Magazine* blog

Ms. blog

HOME   ARTS   GLOBAL   HEALTH   JUSTICE   LIFE   MEDIA   MS.CELLANY   NATIONAL   WORK

You are here: Home / National / The Long History of the War on Contraception

The Long History of the War on Contraception

February 14, 2012 by Ellen Chesler | 5 Comments

For those surprised about the recent fervor over Obama's contraception coverage decision, a look at its deep roots.

Republicans for Planned Parenthood last week issued a call for nominations for the 2012 Barry Goldwater award, an annual prize awarded to a Republican legislator who has acted to protect women's health and rights. Past recipients include Maine Senator Olympia Snowe, who this week endorsed President Obama's solution for insuring full coverage of the cost of contraception without exceptions, even for employees of religiously affiliated institutions. And that may tell us all we need to know about why President Obama has the upper hand in a debate over insurance that congressional Tea Partiers have now widened to include anyone who seeks an exemption.

Fig. 2.28 Barry Goldwater
portrait by Marion Trikosko
for *U.S. News & World Report*

As to epidata, Brian was 17 years old when he made this image of Barry Goldwater. Just a few months earlier he had been sitting in his high school physics class, when the door burst open and the chemistry blurted out between tears that President Kennedy had been shot.

During the election campaign, Brian spent about half an hour with Goldwater and a few campaign workers walking along the main street in the largest city in New Hampshire occasionally chatting with the candidate about politics and about photography—Goldwater was an accomplished, published photographer. This image was first published in the Manchester High School Central school newspaper in the Spring of 1964. At that time there were no crease marks on the image. The original negatives and prints from the shoot disappeared over time. A single small print that Brian had made came to light—quite literally—in 2005 at the bottom of a box of books in his parents' attic. The print had suffered various photo print ravages of time, but a digital copy and some digital editing made it viewable again. The crease marks were left as a reminder of the physical ravages that easily plague analog & chemical photography. The image itself, unremarkable on the face of it, still brings back visceral memories of the times for the photographer.

We contacted the author of the article, Ellen Chesler:

…curious if [you happen to recall why you] used the particular image of Barry Goldwater as [your] illustration, as [Brian] made the photograph in 1964. [we are] well aware that there are more pressing issues all around us, but if a moment arises and the question piques interest, [we] would be most grateful for a reply.

Her response gave substance to our assumption that Flickr.com was simply a convenient source an image that could be used without rights fees:

Actually, it was the folks at Salon, where the piece ran, who chose the image… because the piece mentions Goldwater's support for family planning ….. I suspect a photo editor there just picked it up from elsewhere on the web….(Chesler, 2019)

2.8   Word-Based Documents

Through this journey of various photographs, we see that an image contains (represents) more than meets the eye. Rich, ancillary data that exists around photon data opens avenues of interest. These data, epidata, offer opportunity for proximity as potential points of access increasing the potential of use of the information. Epidata are not limited to images, as evidenced in our opening section discussing Wilson's *Two Kinds of Power*. Epidata presents as a package of data around information painting a picture more vivid in interpretation and subsequent access. Epidata are not one-dimensional offering a single point of proximity. In the culminating section we explore more complex presentations of epidata and proximity.

To this point we have focused our explorations of proximity and epidata on images. Early in our work we briefly mentioned Brian's experience of opening up students' willingness to read Wilson's *Two Kinds of Power*. We turn now to the role of proximity and epidata opening avenues into word-based document discovery. In the following we demonstrate how these notions are much more complex than a few words and often go well beyond the linearity of a single proximal tie.

We opened our narrative sharing the fact that we share a close intellectual relationship. We both are academic descendants of Patrick Wilson. Elfreda Chatman, Laurie's major doctoral studies advisor, studied under Wilson at the same time as Brian, as sketched in Fig. 2.29.

Elfreda served as Laurie's doctoral dissertation advisor for three years, as sketched in Fig. 2.30. Not once during that time did Elfreda mention the name nor did she mention the work of Patrick Wilson.

Fig. 2.29   Sketch of Wilson, Chatman, and O'Connor connections

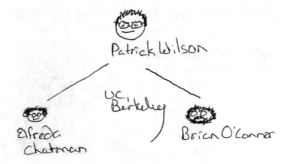

Fig. 2.30 Sketch of Chatman
and Bonnici connection

A year into her career as a professor Laurie accepted a faculty appointment in Texas, but not at the same university where Brian serves as a professor. Their respective universities are located in the same city in Texas. Being in the same field and being in the same city, there is some likelihood our paths would have crossed. However, our meeting actually came about because a mutual friend knew we were both interested in kayaking.

On our first kayaking adventure, as sketched in Fig. 2.31, we discovered our common intellectual connection. The realization coming through a familiarity in language when discussing our research. By language we mean the words we used to describe our research as well as how we expressed our ideas to convey meaning. When one commences out from their academic home of doctoral studies and embarks on the journey of building a career at another institution, this kind of familiarity and intellectual tie is rewarding in both professional work and personal friendships, making for strong ties, as we see in our collaborative work over the years.

Brian mentioned Wilson a number of times during that first kayaking foray, as he continued to do in others over the ensuing weeks and months. Laurie wondered "who is this Wilson fellow?" To Brian's amazement, Laurie had never been introduced to Wilson's work during her time spent under Elfreda's tutelage. Brian once blurted out that he heard

Fig. 2.31 Sketch of Bonnici
and O'Connor connection

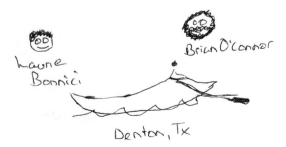

much of Wilson's ideas in Laurie's discussions of research and asked who her dissertation chair had been. This was the step into a web of proximity with Wilson. In Laurie's early explorations of Wilson's work, she was intrigued by *Two Kinds of Power* (Wilson, 1968) as well as *Public Knowledge, Private Ignorance* (Wilson, 1977). But it was *Second-Hand Knowledge* (Wilson, 1983) that she found a common thread connecting Wilsonian thought with her own interests in information sharing in online social media (OSM). Interestingly, Laurie felt that Wilson's work resonated more closely with her own research endeavors than did that of Elfreda's work, as sketched in Fig. 2.32.

Is it possible Elfreda never saw the connection? Whatever the reason, an immediate proximal tie between Elfreda and Laurie evaded an intellectual tie with Wilson's intellectual work. And to further test the potential of proximity, *Second-Hand Knowledge* (Wilson, 1983) is the one work Brian has not read as a published book; he assisted in the indexing, so read it in manuscript. It was the familiarity distinguished through shared language to express research, proximal ties through human relationships (Elfreda and Brian as doctoral student colleagues), Laurie's studying under Elfreda, and a common interest in kayaking shared by Brian and Laurie which resulted in a path of proximity to Laurie's connection with Wilson's work.

Through this ideascape, we present the complexities of proximity and epidata that open avenues to useful information. The notions of shared language, ideas, relationships and

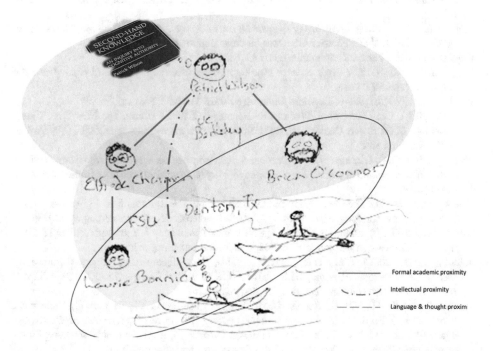

Fig. 2.32 Sketch of complexities of proximity and epidata

hobbies (epidata) make up the information package consisting of epidata (info surrounding info) and the information. Through this first-person journey we realized the complexities evidenced in proximity that make up metadata are not random, but rather form a web of connections and disconnections. We explore webs of connection in more depth in the following sections.

Note

Colorized photographs of Johnson swearing in on Air Force One were produced by submitting a file of the black and white image to each of the following sites, in left to right order

Hotpot AI https://hotpot.ai/colorize-picture

Imagecolorizer.com https://imagecolorizer.com/colorize.html

Cutout Pro https://www.cutout.pro/photo-colorizer-black-and-white/upload.

References

Albannai, T. N. (2016). Conversational use of photographs on Facebook: Modeling visual thinking on social media. Doctoral Dissertation, University of North Texas.

Bonnici, L. J., & O'Connor, B. C. (2019). More than meets the eye: Toward an ontology of proximity. *Proceedings from the Document Academy, 6*(1), Article 12. https://doi.org/10.35492/docam/6/1/13

Chesler, E. (2012). *The long history of the war on contraception.* Ms. Magazine blog. https://msmagazine.com/2012/02/14/conservative-war-on-contraception-is-nothing-new/

Chesler, E. (2019). Personal communication. October 15.

Churchland, P. M. (2012). *Plato's camera: How the physical brain captures a landscape of abstract universals.* The MIT Press.

Farid, H. (2019). *Fake photos. Essential knowledge series.* The MIT Press.

Fox, M. (2008). *Cecil Stoughton Dies at 88: Documented White House.* The New York Times. November 6. Retrieved October 15, 2021 from https://www.nytimes.com/2008/11/06/arts/design/06stoughton.html

Hillard, E. B. (1864). *Last men of the revolution: A photograph of each from life, together with views of their homes printed in colors: Accompanied by brief biographical sketches of the men.* N.A. & R.A. Moore.

Holmes, O. W. (1859). The Stereoscope and the Stereograph. *The Atlantic,* June 1859 issue. https://www.theatlantic.com/magazine/archive/1859/06/the-stereoscope-and-the-stereograph/303361/

Johnson, K. (2011). *Situation: Ambiguous.* The New York Times, May 7. Retrieved October 15, 2021 from https://www.nytimes.com/2011/05/08/weekinreview/08johnson.html

Johnson, L. B. (2001). Diary entry quoted in *Lady Bird*, documentary project produced by McNeil/Lehrer Productions and KLR: Austin, TX. Retrieved October 15, 2021 from https://www.pbs.org/ladybird/epicenter/epicenter_doc_diary.html

Khavin, D., Willis, H., Hill, E., Reneau, N., Jordan, D., Engelbrecht, C., Triebert, C., Cooper, S., Browne, M., & Botti, D. (2021). *Day of rage how Trump supporters took the U.S. capitol.* Investigative Video, The New York Times. Retrieved October 15, 2021 from https://www.nytimes.com/video/us/politics/100000007606996/capitol-riot-trump-supporters.html

Krule, J. (2012). *All the presidents' photographers.* The New Yorker, February 17. Retrieved October 15, 2021 from: https://www.newyorker.com/culture/photo-booth/all-the-presidents-pho tographers

Marr, D. (1982). *Vision: A computational investigation into the human representation and proceedings of visual information.* Freeman.

Mydans, C. (1936). Cigar store Indian, Manchester, New Hampshire. Farm Security Administration collection of the Library of Congress. Digital ID: fsa 8a02862 //hdl.loc.gov/loc.pnp/fsa.8a02862

O'Connor, B. C., O'Connor, M. K., & Abbas, J. M. (1999). User reactions as access mechanism: An exploration based on captions for images. *Journal of the American Society for Information Science, 50*(8), 681–697.

O'Connor, E. M. (2009). Presentation at the Document Academy Annual Gathering, cited in O'Connor, B. C., & Anderson, R. L. (2019). *Video structure meaning.* Morgan & Claypool.

Pelicanwind Photostream on Flickr.com. https://www.flickr.com/photos/pelicanwind/46271141/in/ photolist-2txzi2-569Ni/

Souza, P. (2011). The Situation Room, photograph. Obama White House Archived site on Flickr.com. Retrieved October 15, 2021 from https://www.flickr.com/photos/obamawhitehouse/ 5680724572/

Souza, P. (2020). The Situation Room Photograph, video commentary. Instagram account "pete-souza". Retrieved October 15, 2021 from https://www.instagram.com/tv/B_pZbgllqFv/?utm_sou rce=ig_embed

Stoughton, C. W. (1963). Swearing in aboard Air Force One, photo collection images 1A-1-WH63– 1A-21-WH63. LBJ Presidential Library. Retrieved October 15, 2021 from http://www.lbjlibrary. net/collections/photo-archive.html

Stoughton, C. W. (2002). Cecil W. Stoughton Oral History Interview—JFK#1, 9/18–9/19/2002. Interviewer, Vicki Daitch. John F. Kennedy Presidential Library. Retrieved October 15, 2021 from: https://www.jfklibrary.org/sites/default/files/archives/JFKOH/Stoughton%2C%20Cecil% 20W/JFKOH-CWS-01/JFKOH-CWS-01-TR.pdf

Wilson P. (1968). *Two kinds of power; an essay on bibliographical control.* University of California Press.

Wilson, P. (1977). *Public knowledge, private ignorance: Toward a library and information policy.* Greenwood.

Wilson P. (1983). *Second-hand knowledge: an inquiry into cognitive authority.* Greenwood Press.

Zudark Photostream on Flickr.com. https://www.flickr.com/photos/zudark/4802895581/in/datepo sted/

Epidata, Clues, Threads, Webs

3

3.1 Achieving Proximity

We propose *Epidata* as a term for all the sorts of clues to finding, encountering, and understanding/making use of information that exist outside the ordinarily available tools of data and metadata. Our retelling of the Ariadne and Theseus tale presents the utility of a single clue. Our example of intellectual connections presented in Chap. 2 demonstrate some of the complexities that clues offer in leading to information discovery. Now we sketch some scenarios using multiple clues, forming webs of proximity.

A clue, a ball of thread, whether physical or metaphorical is capable of weaving a conversation with the past and influencing the present. For Theseus, the clue kept track of all the twists and turns of his path through the maze Daedalus had constructed to confine the Minotaur, as sketched in Fig. 3.1, and its victims, then return alive to Ariadne's loving arms from which he had set out.

We recently happened upon *Life Beyond Humans* (Lubow, 2022) which examines artist/scientist Tomas Saraceno's "obsession with balloons and spiders"—a spider web seems to be a useful metaphor/model, as it is a vibrant instrument using subtle clues to alert the spider to events happening within the structure, to potential prey. The clues, the multiple threads in contact with one another and with the spider enable us to sketch the possibly complex relations holding among multiple parties and multiple practices that may hold within a proximity event.

On the first page of the introduction to his *Two Kinds of Power*, information philosopher Patrick Wilson writes: "How can a man be sure of finding, in the great mass of writings, good and bad, pedestrian and extraordinary, the writings that would be of value to him?" On the first page of Chap. 1 Wilson begins his task of making "quite clear what are the different varieties of inhabitants of [the bibliographic universe]" thus:

© The Author(s), under exclusive license to Springer Nature Switzerland AG 2022   53
L. J. Bonnici and B. C. O'Connor, *Proximity and Epidata*,
Synthesis Lectures on Information Concepts, Retrieval, and Services,
https://doi.org/10.1007/978-3-031-17094-2_3

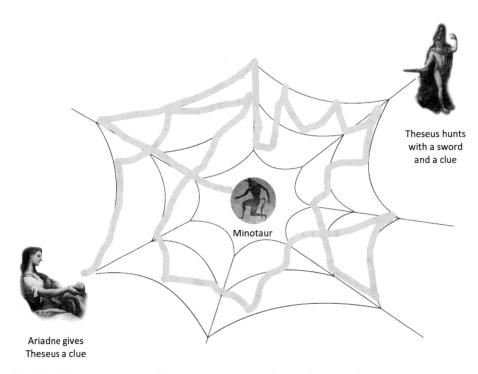

Fig. 3.1 Ariadne's clue keeps Theseus in proximity to the exit from the labyrinth

> A *man* writes a poem, a letter to a friend, a report on an investigation; *he* spends a certain amount of time, a few minutes or many, at consecutive work or spread out over many days, constructing a particular linguistic object, a piece of language. When *he* has finished, that is, when *he* has decided to call the piece of work complete …

In that one paragraph "he" appears eight times. There are some 50 instances of such uses of "man" in the 155 pages of the book; there are 54 instances of person, but none of "woman," so it is easy to see why some contemporary readers would assume some form of gender bias.

It was simple enough to explain to the students who did not want to read Wilson's *Two Kinds of Power* because the author was "sexist," that Wilson, writing in 1968, was operating in a slightly different culture. On the whole, the language was common to Wilson and the students, except for the critical difference of gender-neutral terms; and the interests in bibliography were common, even if Wilson's notions were more philosophical than what the students had previously encountered. There was significant overlap, significant common ground, significant proximity so that a small piece of epidata, a clue, about publisher formatting norms of the day enabled the students to engage with the rest of the work.

A few explanatory threads were enough to weave a web of proximity between the students and the author; the antecedents of both parties were similar and required only a little bit of adjustment.

Let us now turn to some examples of achieving proximity in situations more complex than explaining pronoun usage changes over a thirty-year time span. More complexity than Ariadne giving Theseus a single clue so that no matter his distance from or how many turns he made, he could return to her simply, if not easily. These are situations requiring webs of clues to achieve proximity—webs not just as substrates, but also as a means of maneuvering and evaluating, using the vibrations of the clues together with their stickiness.

3.2   Spinning Webs of Clues: Bounty Hunting and Undercover Police Work

Two pieces of criminal justice research in which we were involved help us to see the finding, spinning, and weaving of clues into a web specifically to achieve physical, Euclidian proximity to a suspect or fugitive. Some years ago, we had opportunities to work with David O'Connor, a highly regarded fugitive recovery agent with an Ivy League degree. At the same time we also worked with Baris Aksakal, an experienced undercover narcotics officer, working on a doctoral dissertation modeling street-level information processing. Over time the two came to know each other and to contribute to each other's thinking. In writing a panel description for an early presentation of this work, we wrote:

> On-the-fly decision making, information juggling, analysis under duress, and the language of daily life come under consideration. Humans are capable of thinking deductively and using deterministic systems, yet they conduct much of their lives thinking analogically and acting on the basis of experience, hunches, and best guesses–greater consideration of these abilities should enhance retrieval system design. (Blair, et al., 2005)

We note that fugitive recovery agent—bounty hunter—O'Connor, said early on that "bounty hunting is an information profession—it is seldom a simple known item search." Similarly, Aksakal, in his dissertation, notes:

> Often the process of information seeking involves the lack of a "known item." This is very true of police work. Even if a description is given of a suspect, this does not lead to the direct location of the criminal. Much work and juggling of information must be done.

The accounts of the search for clues, evaluating them, and weaving them together given by both O'Connor and Aksakal resonate with the description of *evolving searches* in Bates's *berry picking* model:

Each new piece of information they encounter gives them new ideas and directions to follow and, consequently, a new conception of the query. At each stage they are not just modifying the search terms used in order to get a better match for a single query. Rather the query itself (as well as the search terms used) is continually shifting, in part or whole. (1989 p. 408)

3.2.1   What Do I Need to Know in Order to End Up Where the Target Is?

This is the foundational question for tracking down a fugitive, a suspect, or even a document assumed to exist but for which there is no locative data. Someone who has been charged with a crime but wants to stay out of jail until their trial can often post a financial bond to guarantee that they will appear in court on the appointed date. In order to afford the bond, many charged suspects secure loans from bail bond agencies. As part of that process, they are required to give certain diachronic data such as name, date of birth, address, and the like. Thus, they seem to be known items; however, "known item" typically carries the connotation that the 'item's' location is known or easily discoverable. If someone walks into a library and asks: "Where do I find Patrick Wilson's first book?" a librarian can look in a catalogue, find Wilson's books, check the dates on the document records, find the earliest date, look in the location field, and direct the patron to the target document. The readily available diachronic data links directly to a known location. However, the fugitive or the suspect is actively hiding, actively not being not being linked by their diachronic data to a known address. The same could be said to be essentially true for the information a researcher "knows" must exist but cannot describe, provide sufficient diachronic data. This is not to say that the known attributes are useless; they provide initial contact points, possible database search terms, and descriptions for others to use. Here a notice sent to fire stations in locations the fugitive in the case we analyzed might be operating:

On [DATE] I received a call from Dennis Bail Bonds in Chester, New Hampshire. They inquired as to whether I would be interested in taking a contract to find an individual. I stated that I was. I went to Chester to pick up the appropriate paperwork and discuss pertinent particulars. The individual was one Anthony Caruso, DOB 10/05/44. Reason for location was revocation of bail that was posted for the subject at Belknap Superior Court. I was informed that the bail bond was in the amount of $50,000.00 and was issued by a Mr. Ed Brown. Information was relatively sparse:

- Eyes: brown
- Hair: gray
- SS no.: [SOCIAL SECURITY NUMBER]
- Credit ref.: Eileen Caruso (ex-wife)
Tampa, Florida
- Mother: Marguerite Caruso,
Stoneham, Massachusetts
- Sister: Mrs. Robert Jones
[STREET ADDRESS]
Stoneham, Massachusetts
- Brother-in-law: Robert Jones, same address as above
- Girlfriend (?): Claudette Pelton/Pellerin
Rochester/Gonic area?

TO:	FIRE DEPARTMENT PERSONNEL
	FIRE EQUIPMENT SUPPLY COMPANIES
FROM:	XXXX XXXX, Investigator

I am in charge of an investigation to apprehend an individual wanted in several states. A personality trait of this individual is FASCINATION with the Fire Service and associated equipment. He has been observed at the scene of several fires in the New England area impersonating Fire Fighters (especially officers). He has also been seen with fire equipment in his car (hats, bunker jackets, uniforms, etc.). To this point there is NO reason to suspect any involvement in arson. He likes to hang out around fire stations and socialize, usually stating that he was/is involved in the Fire Service in some manner.

This individual has MANY identities and IDs from around the country. He can also convince almost anyone of almost anything. His vital statistics are as follows:

Real name:	XXXX XXXX
Race:	CAUCASIAN
Height:	6'2"
Weight:	PROBABLY AROUND 240 lbs.
Other:	USUALLY HAS WHITE HAIR AND A WHITE CHIN BEARD BUT MAY HAVE FULL BEARD OR BE CLEAN SHAVEN. COMPLEXION MAY BE CRATERED. LARGE SCARS ON STOMACH, TOP OF HEAD, LEFT OF CENTER AT HAIRLINE

If you see this individual, please take NO action yourself. Try to get a license plate number and description of his vehicle. If you are convinced that he is the subject in question and time permits, call your local Police Department and have them check him out. **GIVE THIS LETTER TO THE OFFICERS WHO CHECK HIM OUT.**

I WILL PICK HIM UP

The question: "What do I need to know in order to end up where the target is?" becomes operationalized as "What are the bits and pieces that can be spun into clues that will accurately predict a location for the fugitive and a time that they will be there?" The target attribute is the synchronic, functional attribute: *where* (O'Connor et al., 2003). The search process, the attempt to achieve physical proximity depends on achieving a conceptual, even kinship proximity with the target. Aksakal notes:

> As suggested in the discussion of Chatman's concept of small worlds, undercover police officers have to have an ability to think, speak, act, and even live like the other side – "walk the walk, talk the talk." They have to know the unwritten meanings of words stemming from another culture: they have to be able to play different language games. Wittgenstein gives examples of language games that fit the context of undercover work, such as reporting an event and speculating about an event. Wittgenstein, in addition to Chatman, explains the semantic parts of information transformation that are not so explicit in the Shannon-Weaver model. Both Wittgenstein and Chatman explain the role and context of the language of the undercover world. This language is part of the environment of which the sender and receiver are a part.

We have long used a quote from wildlife photographer and animal tracker, Paul Rezendes (Rezendes, 1992) when speaking of modeling representations of questions and documents:

> Many people today think tracking is simply finding a trail and following it to the animal that made it. …I think the true meaning of reading tracks and signs has in the forest has been

pushed into the background by an overemphasis on finding the next track. …If you spend half an hour finding the next track, you may have learned a lot about finding the next track but not much about the animal. If you spend time learning about the animal and its ways, you may be able to find the next track without looking. Tracking an animal …brings you closer to it in perception.

3.2.2 Serious or Deadly Consequences

Undercover police work differs from fugitive recovery in one significant way—where the bail bond skip is known, the target of an undercover operation may well not be known. That said, the assumptions and processes are largely the same. Aksakal echoes the fugitive recovery agent's comment that bounty hunting is an information profession:

> In the daily lives of undercover police, there are massive amounts of data received from different sources. Inadequate processing of large amounts of data can have serious or deadly consequences. The important question for rookies is how will they separate the needed information from "information pollution"? I searched for a way to put this process into a dynamic, open system structure model of how an undercover cop seeks, gathers, filters, and transforms the information.

3.2.3 Expertise and Trifles

Sir Arthur Conan Doyle (Doyle, 1892) has Sherlock Holmes comment: "It is, of course, a trifle, but there is nothing so important as trifles."Aksakal asserts that in fugitive recovery and undercover police work:

> information seeking and gathering is not just your average subject or keyword search in the library catalogs or even a simple search in police databases. Information seeking in the streets cannot realistically make significant use of the methods that are developed for more conventional information seeking behaviors. (Aksakal, 2005)

- Both formal and informal methods are significant
- Small things matter (as possible clues)
- Expertise is more than the sum of several pieces of knowledge.

The importance of expertise and trifles is emphasized by the Supreme Court case, *Terry v. Ohio* (1968). Aksakal observes that the Court holds:

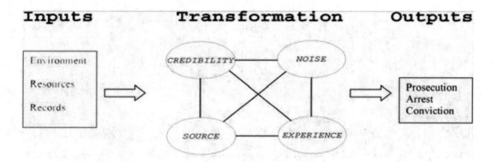

Fig. 3.2   Early conceptualization of Aksakal social virtual interface model

> the nature of police work depends on experience and the ability to comprehend more about conditions associated with criminal behavior than a normal citizen. [H]unches, sixth sense, feelings, variable levels of trust and reliability, fear, pride, and twisting paths are major parts of undercover police work.

Aksakal's model of street level undercover police work, as sketched in Fig. 3.2, assumes "real people are making real information judgments in face-to- face, human-to-human situations." It serves as a "systems model for social network situations that cannot easily be explicitly measured because of the complexity of interactions impacting user behavior," especially when those interactions are woven "together with the very real danger involved in these situations."

Dreyfus and Dreyfus assert that experts gain their expertise with "training through numerous practice and "real" situations," (Dreyfus & Dreyfus, 1986), giving credibility to the notion of police expertise or "sixth sense" recognized by the Supreme Court's decision *Terry v. Ohio* (U.S. Supreme Court, 1968). Aksakal's analysis of the descriptive statistics for the transformation stage indicators of the model, together with content analysis of interviews with veteran undercover police indicate that the subjects "believe that their job requires more hunting than just simple gathering [and that it] requires more experience and adaptation to the dynamic characteristics of those environments."

Aksakal's early notes and sketches of non-linear and poly-dimensional qualities of "street-level" police work, as shown in the sample in Fig. 3.3, uses "the sizes, the irregular lines, and the overwriting [to] demonstrate the multidimensional and multilayered nature of the model better than standard graphical representations." (Aksakal, 2005, p. 30).

Speaking of Aksakal's model of the "general, non-linear, not 'perfect,' and 'make-shift' nature of information flow and processing in street-level undercover police work" David O'Connor notes, "[I]t better defines the complexity of the problem and better illustrates where many of the contentious issues originate. But there also was a great need within the police community itself for a model that clearly defined how the "process" of effective

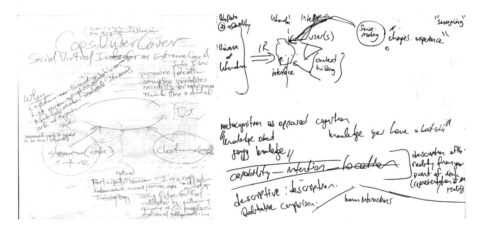

Fig. 3.3 Early sketch of Aksakal non-linear model of "street-level" police work

undercover police work proceeded. And the multitude of factors that can effect – either positively or negatively – that process." (Personal communication, April 22, 2005).

3.2.4 Fifty-Two Stories to an Arrest

In an earlier publication (O'Connor et al., 2003) we presented a 38-page condensed version of the daily notes kept during a six-month investigation, commenting: "it is only with such thick description that we see the importance of small details, the importance of persistence, the importance of personal commitment, and the importance of analyzing failures for clues to new paths." We had titled the chapter in which we discussed this case *52 Stories to an Arrest* because there were 52 primary components to spinning the threads of the clues; they are summarized as:

24 people—relatives, acquaintances, victims
4 hotels/motels
4 car rental sites
3 credit card companies
State trooper
Motor Vehicle Department
Fire Departments
Parcel shipping company
Muffler repair shop
Realtor
Picture and flyer distribution

Local law enforcement

National chain department store

Examine trash

Who might be persuaded to give substantial information?

Our content analysis of the activities gives another view of the spinning of the threads:

Analyzing	Following
Strange telephone call	Fugitive's mother
Arresting	Fugitive's relative
Fugitive	**Inquiring**
Calling	Police department
Bail bond company (6 times)	Post office
Building owner	**Locating**
Car rental agencies (various—8 times)	Possible girlfriend
County sheriff	**Following**
Credit card companies (9 times)	Fugitive's mother
Ex-wife of fugitive	Fugitive's relative
Father of woman involved	**Inquiring**
Fire marshal	Police department
Fire services (numerous)	Post office
Hotel	**Locating**
Karen	Possible girlfriend
Midas (2 times—fugitive seen)	**Mailing**
Motel (3 times—confirm use of stolen card; fugitive seen)	Flyers (wanted notices—2 times)
New York City Fire Service	**Meeting**
New York jeweler (3 times)	Robert (3 times)
Owner of stolen wallet	**Observing**
Possible contact (3 times)	Coast
Relative of fugitive	Fugitive's mother (2 times)
Robert (11 times)	House (2 times)
Sawyers	**Reviewing**
Son of fugitive's wife (7 times)	Notes
Source (3 times)	**Running**
State trooper (2 times)	Credit card checks (numerous)
Supposed employer	License plate number

Tampa sheriff	Vehicle
Telephone company information	**Talking**
Telephone number on rental slip	Bail bond agency
UPS security (4 times)	Contact
Checking	Karen (2 times)
Address	Possible lead
Credit card use (2 times)	Robert (4 times)
Fugitive's aliases	Sawyers
License plate numbers (numerous)	Trooper
National criminal database	UPS security
Contacting	**Using**
Bail bond agency	Public library
Brother of fugitive's wife	**Visiting**
Building owner	Address in strange call
Car rental manager	All motels in 10-mile radius
Fire and police departments (numerous)	Bob and Karen
Florida bail bond agency	Burlington (wanted notices)
Son of fugitive's wife (2 times)	Logan Airport (wanted notices)
Telephone company	Norway, Maine (wanted notices)
UPS security	Police departments
Discussing (with)	Relative
Bail bond agency (2 times)	Robert
Sheriff's department	Sears (interview staff)
Distributing	
Flyers (wanted notices)	
Evaluating	
Evidence	
Situation	
Faxing	
Police (arrest arrangements)	
UPS	

Our analysis of the case report and subsequent discussions with the recovery agent yielded a general set of activities for spinning clues and weaving a web for recovery of fugitives:

Fig. 3.4 Outline sketch of fugitive recovery

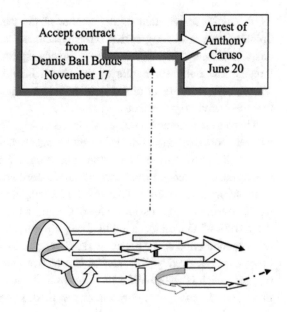

- generate several threads or lines of investigation
- establish, monitor, maintain, generate collaborators
- monitor all the lines of investigation
- evaluate progress
- generate new threads
- modify threads
- abandon threads
- look for anomalies
- inform the generation and monitoring of the threads
  - previous experience and
  - knowledge of what fugitives do.

At that time, we constructed a multi-layered model reflecting the variety of tactics, number of collaborations, and constant juggling and evaluation represented in the attempt to construct a web of proximity, summarized in Fig. 3.4 and the following.

3.2.5 Functional, Pragmatic, Contingent, Satisficing Model

We established the model on Copeland's "functional, pragmatic, contingent, and satisficing" model of engineering design activity (O'Connor et al., 2003). We elaborated on the model with callouts to collaboration, spinning and evaluating search threads, and the nonlinear nature of the process. Collaboration included assistance in one event, team member for the entire search, temporary member of the team, evaluation of each collaboration. Spinning and evaluating search threads included launching multiple threads simultaneously, evaluating threads, discarding threads, spinning threads together, launching new threads. The nonlinear nature of the search included frequent evaluation of the entire search, acknowledgment of the iterative nature of portions of the process, and acceptance that the search might arduous.

These search components, as sketched in Fig. 3.5 fit our analysis of the case study and our conversations with the recovery agent; however, the illustration of the model was too linear to represent their interconnections. We sought a modeling structure similar to Copeland's bricoleur model in Fig. 3.6, a nonlinear model incorporating the pragmatic and contingent necessities of street-level investigative work. Those connections were highlighted during our discussions with O'Connor and Aksakal, when we brought up Bates's "idea tactics." O'Connor noted that they are quite similar to the search tactics in fugitive recovery and undercover police work. He emphasized that they are not deployed sequentially but concurrently. The idea/search tactics and the threads of the investigation are juggled—each is analyzed for its progress and potential, then many are evaluated in relation to others, continued, dropped, or modified. Expertise, trifles, and evaluation yield the epidata, the web of clues in both the ancient sense and the contemporary sense, that bring about physical proximity with the target. We adapted our analyses and the models from Aksakal, Copeland, and O'Connor to the web of proximity format as in Fig. 3.7.

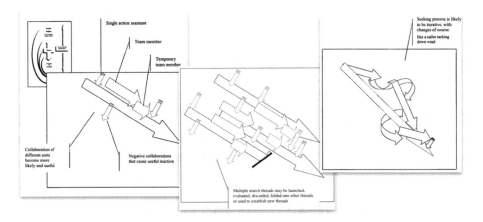

Fig. 3.5 Composite model of searching for unknown targets, adapted from O'Connor (O'Connor et al., 2003)

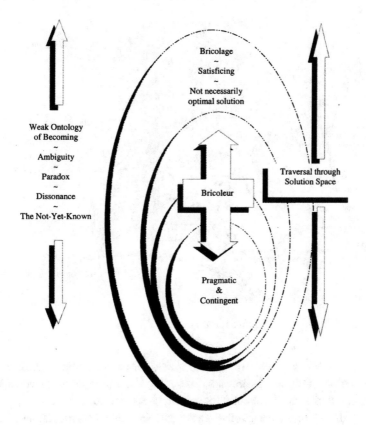

Fig. 3.6   Model of engineering design activity, adapted from Copeland (O'Connor et al., 2003)

3.3   Shaking Up the Personal Knowledge Store: Browsing and Idiocy

We turn here to scholarly browsing as a search activity with no specified target and no initial clues. Academic scholars are charged with being creative, with generating new knowledge. Knowledge yet to be generated cannot yet have been classified or indexed. That is, one cannot walk into a library or login to a search engine and search with descriptors for something that does not yet exist, that has not yet been described. One cannot walk into a library or logon to a digital platform and ask for "my new knowledge paper" or even "existing works likely to spark a new idea." How is a scholar to achieve proximity to a documentary source when neither the question nor the likely source of satisfaction is known?

Browsing has long been a response to this question. Browsing is fundamentally a matter of serial proximity to document after document until one is found that works, that inspires, that fills in the missing piece. It is notable that "browse" refers to animals feeding

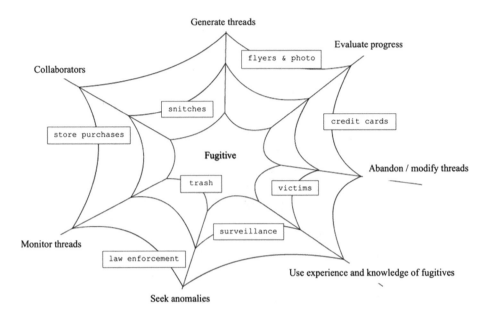

Fig. 3.7 Web of proximity for fugitive tracking

in the wild, "sometimes carelessly used for *graze,* but properly implying the cropping of scanty vegetation." (Weiner & Simpson, 2004) The sheer number of documents on stack shelves or in a digital environment does not imply creative sustenance.

We use "idiocy" to emphasize the utterly personal, idiosyncratic nature of browsing. Idiocy might seem a harsh term, yet it relates directly to creativity and to navigation for creative purposes. While its current use is essentially pejorative, its etymology fits the notion of creativity. Liddell and Scott tell us that the Greek root—idios—means:

ἴδιος
I. *one's own, pertaining to oneself*: hence,
1. *private, personal,*
2. *one's own,*
3. τὰ ἴ. *private interests,*
II. *separate, distinct,*
b. ἴ. λόγος, in Ptolemaic and Roman Egypt, *private account,*
2. *strange, unusual*; *peculiar, exceptional, eccentric,*
3. *peculiar, appropriate* (Liddell & Scott, 1940).

3.3.1   Bring Forth, Produce, Beget, Cause

Creativity has at its root "to grow;" beyond that root there is a great deal of speculation, debate, and a considerable number of experimental inquiries. Sculptor George Segal has used "total goulash" to describe the complex web of attributes that comprise this state/set of states. (Isakson, in O'Connor, 1988) Some general characterizations of creativity seem to hold. For our examination of proximity, we can say it is "a unique integration of various mental operations," (Haley, in O'Connor, 1988) in large part an intentional, and conscious attempt to reconcile or reformulate seemingly antithetical elements into a new concept embodying the 'truth' of each of the seminal elements. (Rothenberg, in O'Connor, 1988) Such reconciliation or reformulation requires the ability to 'know thyself' (Demey, in O'Connor, 1988), that is, to be in a 'state of readiness for catching similarities' and to be 'gullible', making free-associations. (Arieti, in O'Connor, 1988) All manner of connections are made willingly, especially those which seem 'illogical' or unsanctioned by current knowledge or models; yet the connections remain subject to critical evaluation.

3.3.2   Locus of Representation

We assert that searching for information/inspiration in a document environment is an attempt by the searcher to bring their knowledge state attributes surrounding unknown values (reduced here to a small array for simplicity) into proximity with similar arrays in other sources—documents representing knowledge states of their authors. We might say that one is aware of one's own antecedents (e.g.: language, topic knowledge, time frame, experiences, relationships) and is seeking similar bundles of antecedents wrapped around what is missing. A document that has some threshold of matching antecedents and a value in the cell that is empty in the seeker's array has some probability of being useful. Search systems generally establish forms of representation as well as general rules for how questions can be put to the system.

We can model the document search as an attempt to bring the seeker's knowledge state into a class of documents, as in Fig. 3.8. Simple classification theory tells us that members of a class share all (classical) or some significant threshold (probabilistic) of attributes; so, if we can find a class that matches some or all of the scholarly searcher's known attributes with one or more documents with the same attributes, then the blank attributes in the scholar's array, may be filled by values in the corresponding cells in the documents in the class. Bringing the scholar's array into proximity with the document arrays may well provide satisfaction.

What though of the case of the browsing scholar? essentially all attributes of the scholar, all antecedents, are at play, as are all the attributes of all the documents. A scholar navigates a document collection in hope of a 'eureka' situation—the discovery of new knowledge, of new syntheses. Such seeking often relies on making new connections

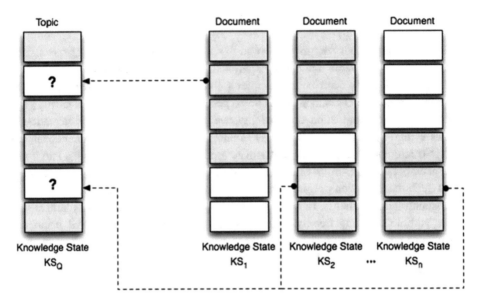

Fig. 3.8 Searchers and documents represented as knowledge state arrays for classifying

between attributes of documents, between concepts in the user's knowledge store (with document attributes as catalyst), and between the user's concepts and attributes of documents. The scholar seeking at the frontiers of knowledge requires an approach which is not anchored to existing knowledge and relationships; they must turn to inward; they must shift the locus of representation.

In a very small collection or in some ideal world, it would be reasonable to say: "Look at every document in the collection and if there is anything useful there you will find it." For searches in which there is some idea of what would be a suitable document standard representation schemes can achieve useful degrees of proximity; they carve the search space into manageable chunks. In the scholarly search there are few means of carving the search space to manageable chunks, save for non-topical attributes such as documents in languages not spoken/read, fields in which the searcher is already knowledgeable (Fig. 3.9).

Browsing generally relies on some form of sampling to carve the search space to a manageable size, to accomplish within real world constraints the possibilities of an 'ideal' retrieval system which would screen all user attributes against all document attributes of all works in the collection (Robertson et al., in O'Connor, 1988). It is a matter of sampling texts at various levels of penetration into the collection as a whole and into any individual document in order to make a judgement as to whether a document has sparked a eureka moment, then to decide whether or not to move on to sample more (O'Connor, 1988). The seemingly serendipitous discovery is not simply a matter of blind luck, rather it is the recognition of a valuable document attribute/connection discovered by means outside

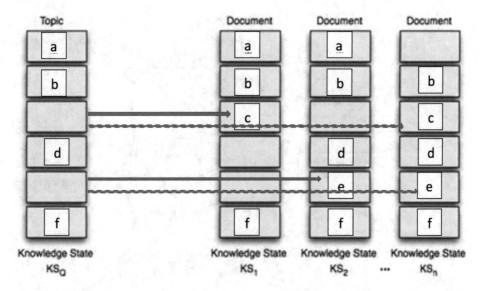

Fig. 3.9 Browsing as sampling in documents sharing some attributes with seeker in hopes of filling gaps or connecting concepts when the gaps are unknown

established access system rules and relying on the user's self-knowledge. We have noted in earlier work that browsing both burdens and empowers the scholar by:

- Enabling control of location and depth of penetration in a collection
- Formulating the rules of representation—what is highlighted and what left behind
- Determining the acceptability of tradeoffs in representation
- Determining the sampling method.

The web of proximity a eureka moment in Fig. 3.10 for relies on the browsing scholar's ability to sample and to glimpse at data points in documents at loci and depths determined by themselves and on the fly as the search progresses. Little else is at hand. Document attributes are sampled and evaluated; strands of clues emerge, some to be seen no more and some yield eureka.

Let us look at the weaving of a complex web closely related to the Wilson gender-neutral pronoun case. We will present a good deal of detail in order to emphasize the scale of the weaving project.

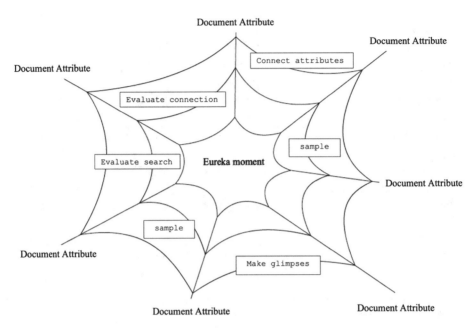

Fig. 3.10 Web of proximity for browsing and creativity

3.4   ἄνδρα μοι ἔννεπε, μοῦσα, πολύτροπον

The case of Wilson's *Two Kinds of Power* was simple, in the sense that both parties—the author and the readers—used essentially the same language and were engaged in the same topic. How, though, is a contemporary English reading audience to find proximity with this message?

ἄνδρα μοι ἔννεπε, μοῦσα, πολύτροπον, ὃς μάλα πολλὰ

πλάγχθη, ἐπεὶ Τροίης ἱερὸν πτολίεθρον ἔπερσεν

Many of the characters look like those of the modified Roman alphabet we use, many do not. We have shown this message to numerous people over the years and a frequent (correct) response is: "It's Greek to me." Only very seldom has this meant the person knew Greek; typically, it has been a way of saying: "It makes no sense to me." Wyatt Mason interviewing Odyssey translator Emily Wilson notes he said to her: "Treat me," I interrupted, 'as if I don't know Greek,' as, in fact, I do not." (Mason, 2017).

How are we to achieve some form of functional proximity in a case where the separation between author and reader is not 30 years difference in a common language, but 3000 years in a very different language? *The Oxford English Dictionary* reminds us that the earlier definition of proximity was: The fact or condition of being near or close in abstract relations, as kinship (esp. in **proximity of blood**), time, nature, etc.; closeness.

Since this text is one of the oldest in the Western literary tradition, there are translations. However, translations have shortcomings. Patrick Wilson notes:

> A translation must preserve the sense of its original, and it is not hard to imagine, or find, instances of texts claiming to be translations of such and such a work, but bearing so little resemblance to the original, preserving so little of the sense of the original as to be 'no trans-lation at all.' But there is no imaginable way of saying precisely how much of the sense of the original must be preserved, for a putative translation really to **be** a translation of some text. (Wilson, 1968)

Scott notes in the translators' introduction to Plato's *The Republic*, the proximity issues in maintaining "the sense of the original" and suggests "conversation" as an approach:

> Frustration results from the recognition that translation can never reproduce the original. Vari-ances in vocabulary, sentence structure, word order, idiom, and cultural context are not only characteristics that make every language unique; they are also the principal reasons why translation must be a matter of conversation and not mere copying (Plato et al., 1996).

Klaver enhances the notion of translation as conversation, as a weaving of proximity:

> Translation is the beginning of understanding, the shadow of thought, inviting further think-ing. Translation reveals a transitional quality of thinking. It facilitates transitions, connections, relations, opening new ways of understanding, shedding new light on situations. It is predi-cated upon an ontology of being with, an epistemology of knowing *with*. (Klaver, 2018)

A translation of a text such Homer's *Odyssey*—from which the passage above is derived—requires more than a Greek—to—some modern language dictionary precisely because of Scott's "variances." A translation is an attempt to achieve proximity by reducing the opacity of the variances. The processes and environment of translation can be seen as a conversation between a translator, a text, and the antecedents of the author of the text; together with a companion conversation between the translator, an audience, and their antecedents. A translator spins clues from specialist understanding of languages and cul-tures and requirements then weaves a web of those clues, a web of epidata between the ancient author's antecedents and those of the modern audience.

While 'translation' of Patrick Wilson's *Two Kinds of Power* for an audience three decades later required only a few minutes effort and brief discussion of a few social and linguistic differences, translating a text from a distant time and place, such as Homer, requires the spinning of many threads and weaving an extensive conversation with mul-tiple parties (Fig. 3.11). To come to a sense of how the clues are spun then woven into a web of proximity, we look first to Mason's thoughtful and extensive review of Emily Wilson's translation, then to Wilson's own notes.

Fig. 3.11 Homeric papyrus of the *Odyssey*, Book 10, lines 527–556. Late second century A.D. University of Michigan Papyrology Collection

There are more than six dozen published English translations of the *Odyssey,* beginning with Chapman's in 1616; There are 13 published in the first two decades of the twenty-first century. Emily Wilson's 2018 translation is the subject of our final sketch. Wilson's conversation with the antecedents is extensive.

3.4.1   Tell Me About a Complicated Man

Reviews for Wilson's translation of Homer's *Odyssey* bespeak her ability to achieve a web of proximity to Homer's ancient world for the modern reader. *Library Journal* notes: "Wilson offers a fluent, straightforward, and **accessible** version." Acclaimed classicist Mary Beard writes: "Wilson's version is an exciting one …forcing us to think how **approachable**, how weird, how **everyday** Homeric language ever was." The *Financial Times* says:

"Wilson's translation feels like a restoration ...she **scrapes away at old encrusted layers**, until she exposes what lies beneath."

Wilson summarizes her approach to the translation, echoing Scott's variances, thus:

> I have taken very seriously the task of understanding the language of the original text as deeply as I can, and working through what Homer may have meant in archaic and classical Greece. I have also taken seriously the task of creating a new and coherent English text, which conveys something of that understanding within an entirely different cultural context.

She also notes:

> "My Homer does not speak in your grandparents' English... I have tried to keep to a register that is recognizably speakable and readable...". (Wilson, p. 87)

It requires only a small sample of the English translations to demonstrate the variances existing between just those translations; the differences in times and cultures yielding differences in the words not so opaque as the Greek, yet not necessarily transparent.

Four translations of the first lines of *The Odyssey*

The Man, O Muse, informe, that many a way

Wound with his wisdome to his wished stay;

(Chapman, 1614 in Nicoll, 1965).

Tell me, O Muse, of that many-sided hero

who traveled far and wide.

(Butler, 1900).

Tell me, O Muse, of the man of many devices,

who wandered full many ways

(Murray, 1919 in Homer et al 1995).

Tell me about a complicated man.

Muse, tell me how he wandered and was lost

(Wilson, 2018).

The fifth word in the very first line of the Greek poem is then used to illustrate this point. The adjective πολύτροπον—polytropon—is used to describe the central figure of the tale well before we are given the hero's name, so we might assume it is of some considerable importance in establishing the man's character/fate. Yet, among the 60 predecessors to Wilson there are more than forty translations of that word, ranging from "crafty" to "tost to anf fro by fate," from "skilled in expedients" to "shifty," and from "Ingenious" to "for shrewdness famed/and genius versatile." Most opt for straightforward assertions of Odysseus's nature, descriptions running from the positive (crafty, sagacious, versatile) to the negative (shifty, restless, cunning)." The Liddell & Scott Greek to English dictionary renders polytropos thus:

πολύ-τροπος
A. *much-turned*, i.e. *much-travelled, much-wandering*, epith. of Odysseus
    II. *turning many ways*: metaph., *shifty, versatile, wily*
2. *fickle.*
3. of diseases, *changeful, complicated,*
       III. *various, manifold.* (Liddell & Scott, 1940)

Wilson notes that Odysseus "seems to contain multitudes: he is a migrant, a pirate, a carpenter, a king, an athlete, a beggar, a husband, a lover, a father, a son, a fighter, a liar, a leader, and a thief" (p. 5) and that "Most of the epithets applied to Odysseus begin with the prefix *poly-* meaning 'much' or 'many': he is a s figure who possesses many attributes and possesses them intensely." Yet, he is beset by so many trials, distractions, calamities, and challenges—so many *turns* that he spends 10 years simply trying to makes his way back home from the Trojan war. To accommodate this duality, this confusion of Odysseus being battered by turns of fortune and Odysseus turning matters his way that πολύτροπον foretells, Wilson translates it as "complicated," as we see in the Liddell and Scott usage for describing a "disease."

Mason (2017) quotes Wilson on using *complicated*:

> I wanted there to be a sense [that] maybe there is something wrong with this guy. You want to have a sense of anxiety about this character, and that there are going to be layers we see unfolded. We don't quite know what the layers are yet. So I wanted the reader to be told: be on the lookout for a text that's not going to be interpretively straightforward.

3.4.2   Clues and a Web of Proximity

Wilson asserts that this "very ancient and very foreign text" presents "concerns such as "loyalty, families, migrants, consumerism, violence, war, poverty, identity, rhetoric, and lies [that] are in many ways deeply familiar, but we see them here in unfamiliar guises." How might we model the hope that "my translation will enable contemporary readers to welcome and host this foreign poem, with all the right degrees of warmth, curiosity, openness, and suspicion"? What are the clues Wilson weaves into a web of proximity connecting the modern English-speaking reader?

We propose two sorts of clues, skeins of thread, epidata: those which connect the antecedents of the author of "very ancient and very foreign text" with the antecedents of the "contemporary readers." The antecedents are of two related sorts: the conventions for observation and action of the ancient author and of the contemporary readers (what might be regarded as the abstract universals of each) and the conventions for representation of each. Wilson's introduction provides a scaffold of antecedents that include, though are not limited to:

Conventions of observation and action in the time of Homer

- Oral tradition: patchwork of dialects, times, places
- Geographical setting: hard to pin down, some real, some fantastical
- Hospitality & guest-friendship: distinctive bond, episodes of violation
- Gods: like humans
- Women: chattel, fidelity, portrayed sympathetically
- Becoming a man: role of the son of Odysseus
- Slaves: fact, tools of the elite, major role.

Conventions of representation in *The Odyssey*

- Predictable rhythm of orality
- Formulaic expressions loosened
- Wide range of stylistic registers
- "…stylistic pomposity is entirely un-Homeric
- Vocabulary stripped of modern agendas
- Contemporary English very different
- Sexism and patriarchy: within.

To facilitate proximity, to reduce the impact of variances, to engage the reader in the conversation, Wilson wraps the 12,110 lines of poetry in 80 pages of introductory material, 11 pages of translator's notes, 6 pages of maps, 26 pages of notes, 26 pages of a glossary, and a Twitter feed. The last of these brings into play yet another instrument to weaving a web of proximity for the "contemporary reader." The aptly termed "Collected tweet threads" link opens with these words: "Professor Emily Wilson's Twitter feed (@EmilyRCWilson) provides her readers with insights into the art of translation."

The radial threads (in red) are the antecedents of both the original author and the current recipient of the message, the conventions for observation and action, together with the conventions for representation. The concentric threads are the tools and processes by which the translator constructs a translation conversation. We use "translation conversation" to emphasize that achieving proximity requires reducing the impact of "[v]ariances in vocabulary, sentence structure, word order, idiom, and cultural context" (Sterling, 1985) and that understanding requires an "ontology of with, an epistemology of *with*." (Klaver, 2018)

The proximity web in Fig. 3.12 is the affordance, the field of epidata that enables understanding. The document has already been found, its physical proximity is close, literally at hand. The epidata, whether the tiny bit of information about publishing practices when Patrick Wilson wrote *Two Kinds of Power* in 1968 or the years of work and expertise of Emily Wilson weaving a conversation for a contemporary audience of Homer, affords a conceptual proximity between distant partners in a conversation engaging in a dance of understanding.

The complexity of Homer and the attempts at translation we offer here demonstrate some of the most formal, academic efforts to approximate the user with a document. The translation context allowed us to tease out and then demonstrate our notion of epidata as an often complex web of antecedents and connections. But what of less formal, more everyday encounters where epidata lead us to useful information. In the remaining section we offer a series of provocations extrapolated from our own experiences. Within each we pose questions intended to provoke further thinking and application of epidata in the information seeking process.

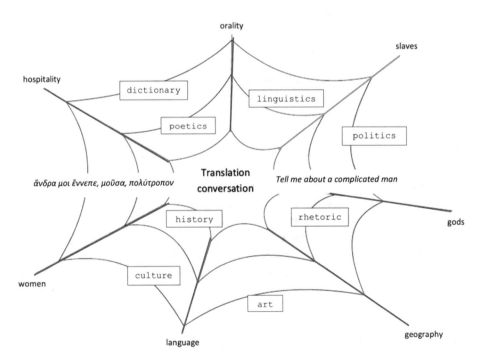

Fig. 3.12 Simple model of web of proximity achieved through translation conversation

Our explorations of Webs of Proximity opened with the tale of Theseus and Ariadne. The pursuit of the Minotaur by Theseus was simple, escaping the labyrinth was thought to be impossible. Ariadne's offer of a skein (klewe, clue) of yarn that would maintain contact between her and Theseus as he journeyed into the labyrinth is an exemplar thread in the Web of Proximity. This scenario is the evidence of myth, so we presented instances of epidata in our real-life encounters that hold potentially impactful clues for our discipline and others. We presented instances based on photographs, cases based on word-based documents, and cases built, literally, on hard evidence to weave an illustrative web of proximity constructs.

Next, we present sketches of other encounters with epidata, along with some questions. In essence, we offer these provocations to ignite ideas of how webs of proximity present epidata that, although not centrally informative, present possibilities of finding or stumbling upon information, or hold potential to add value to information facilitating a seeker's decision to engage with it, to find it useful.

References

Aksakal, B. (2005). Makeshift information constructions: Information flow and undercover police. Ph.D. dissertation, University of North Texas. https://digital.library.unt.edu/ark:/67531/metadc 4823/. Accessed June 13, 2022

Bates, M. J. (1989). The design of browsing and berrypicking techniques for the online search interface. *Online Review, 13*(5), 407–424. https://doi.org/10.1108/eb024320

Blair, D., O'Connor, B., Bonnici, L., Chilton, B., & Aksakal, B. (2005). Perspectives of information seeking and gathering behavior in high-risk professions, panel session. *Proceedings of the American Society for Information Science and Technology.* https://doi.org/10.1002/meet.145041 0184

Doyle, A. C. (1892). *Adventures of Sherlock Holmes, vi The Man with the Twisted Lip.* George Newnes Ltd.

Dreyfus, H., & Dreyfus, S. (1986). *Mind over machine.* Macmillan.

Homer, & Butler, S. (1900). Gutenberg Project eBook produced by Tinsley, J., & Widger, S. (1999). https://www.gutenberg.org/files/1727/1727-h/1727-h.htm

Homer, Murray, A. T., & Dimock, G. E. (1995). *The Odyssey.* Harvard University Press.

Homer, Wilson, E. R., & Homer. (2018). *The Odyssey.* W.W. Norton & Company.

Homer's Odysses Translated according to ye Greeke by Geo: Chapman, in 1956 Chapman's Homer: The Iliad, the Odyssey and the Lesser Homerica (Vol. 2). Edited, with Introductions, Textual Notes, Commentaries, and Glossaries, by Allardyce Nicoll. Bollingen Series XLI. Pantheon Books. https://archive.org/details/chapmanshomerili02home/page/n9/mode/2up

Klaver, I. J. (2018). Trans-scapes transitions in transit. In R. Scapp, & B. Seitz (Eds.), *Philosophy, travel, and place: Being in transit.* Palgrave Macmillan.

Liddell, H. G., & Scott, R. (1940). A Greek-English Lexicon. Revised and augmented throughout by Sir Henry Stuart Jones with the assistance of Roderick McKenzie. Clarendon Press. http://www.perseus.tufts.edu/hopper/text?doc=Perseus%3atext%3a1999.04.0057

Lubow, A. (2022). The world's most amazing Spider Man. New York Times Style Magazine, February 9, 2022. https://www.nytimes.com/2022/02/07/t-magazine/tomas-saraceno-spiders-shed.html

Mason, W. (2017). The first woman to translate the Odyssey into English. New York Times Magazine. NYT November 2, 2017. https://www.nytimes.com/2017/11/02/magazine/the-first-woman-to-translate-the-odyssey-into-english.html

O'Connor, B. (1988). Fostering creativity: Enhancing the browsing environment. *International Journal of Information Management, 8*, 203–210.

O'Connor, B. C., Copeland, J. H., & Kearns, J. (2003). *Hunting and gathering on the information savanna: Conversations on modeling human search abilities.* Scarecrow Press.

Plato, Sterling, R. W., & Scott, W. C. (1996). *The Republic.* Norton.

Rezendes, P. (1992). *Tracking and the art of seeing: How to read animal tracks and signs.* Camden Housed.

U.S. Supreme Court. (1968). Terry v. Ohio, 392 U.S. 1 No. 67. https://supreme.justia.com/cases/federal/us/392/1/

Weiner, E. S. C., Simpson, J. A., & Oxford University Press. (2004). *The Oxford English dictionary.* Clarendon Press.

Wilson, P. (1968). *Two kinds of power: An essay on bibliographical control.* University of California Press.

Provocations and Invitations

4

4.1   Stumbling Upon a Clue

Why does Laurie's neighbor have a small private collection of official US Presidential photos?

#clues
#archives
#UShistory
#epidata

Our explorations of Webs of Proximity opened with the tale of Theseus and Ariadne. The pursuit of the Minotaur by Theseus was simple, escaping the labyrinth was thought to be impossible. Ariadne's offer of a skein (klewe, clue) of yarn that would maintain contact between her and Theseus as he journeyed into the labyrinth is an exemplar thread in the Web of Proximity. This scenario is the evidence of myth, so we presented instances of epidata in our real-life encounters that hold potentially impactful clues for our discipline and others. Here we present sketches of other encounters with epidata, along with some questions. In essence, we offer these provocations to ignite ideas of how webs of proximity present epidata that, although not centrally informative, present possibilities of finding or stumbling upon information, or hold potential to add value to information facilitating a seeker's decision to engage with it, to find it useful.

Recently I was asked by a neighbor, who had a day of scheduled errands, to look in on her elderly pet who had been feeling poorly. I had had no previous close contact with my neighbor, Judy, other than frequent passing encounters during neighborhood forays with our dogs. My only knowledge about her was that she hailed from Washington, DC, several hundred miles away from where we both currently live.

© The Author(s), under exclusive license to Springer Nature Switzerland AG 2022   79
L. J. Bonnici and B. C. O'Connor, *Proximity and Epidata*,
Synthesis Lectures on Information Concepts, Retrieval, and Services,
https://doi.org/10.1007/978-3-031-17094-2_4

Fig. 4.1 Three original prints of presidential photos in neighbor's home

When I went to my neighbor's home and opened the door, Judy's dog Lily was nowhere in sight, so I ventured into the abode. Lily either had significant hearing loss or was choosing to ignore my summons. I went in search of her throughout the home. Only in passing did I notice any elements of décor or design. With luck, Lily popped up from a chair, half hidden under a blanket, located under a small collection of black and white photos hanging on the wall. As I greeted the aging pup, I noticed that the photos included Lyndon Johnson as a subject (Fig. 4.1). My interest was piqued, but Lily's desire to go outside trumped further viewing. While slowly exploring the neighborhood with Lily I found myself wondering about the photos I just saw. While musing, I recalled that every US president starting with Kennedy had an appointed White House photographer. My knowledge informed by a recent co-authored publication centered on an informational exploration of presidential photos (Bonnici & O'Connor, 2021).

Upon my return from accompanying Lily on her constitutional, I took the opportunity to explore the framed photos on the wall more closely. Sure enough, I noted that each of the three photos was signed by Yoichi Okamoto, Johnson's official photographer. Given the quality of the photos and the signatures on the matting, I knew that these must be authentic pieces. Given the era these photos document, they did not correlate with my neighbors speculated age. Thus, I found a cognitive disconnect in the relationship between photos and owner. To me, Judy appeared to be too young to have worked in DC during the Johnson years. How did she acquire items of such limited availability? How could I not ask?

The next evening, I ran into Judy as we walked our pups in the neighborhood. After a brief exchange of appreciation for helping Lily the day before, I mentioned that I had noticed the presidential photos on her wall. I reaffirmed that she had worked in DC, but that I found it intriguing as she was much too young to have served in Washington during the Johnson era. A hearty laugh and an assurance affirmed my thinking. Then she followed with "well, that's an interesting story of how I got those." The story goes like this…

4.1.1 #Clues

Judy had worked in DC during the Reagan years as a lobbyist. During her career she had become quite knowledgeable of people and processes, as is often the case with DC-based employment. About a year after she retired, a friend contacted her inquiring about her knowledge about some aspect work he was attempting to accomplish involving valuation of collectibles. She passed on to him name and contact info of someone in DC she thought could help. A few months later he asked Judy to lunch when she was making a visit to DC. He wished to thank her for connecting him with someone who helped him significantly with his work. He passed her a portfolio of photos from the Johnson photographer's trust collection that were gifted to him and invited her to choose any three she wished to have.

She chose the three I had noticed hanging on her wall. But why those photos? Judy shared with me that the Johnson years were a part of her formative years. He served as president at a time when she was young, impressionable, and had strong ideas. She appreciated that these photos were connected with significant events in her lifetime.

4.1.2 #UShistory

Image 1

According to Judy, Image 1, shown in Fig. 4.2, sparked an important memory in her life. Eisenhower was the first election she experienced as a child and she remembers listening to the outcome on the radio with her father. Thus, the photo provided a tangible connection/clue to a memory of a cherished loved-one. More specifically, the image documents a vacant time in US history. Johnson became president upon the assassination of John F. Kennedy. The vice-presidency position sat vacant for a time. A gap in presidential service, due to Kennedy's demise, made for a unique proximal connection between Eisenhower and Johnson in the line of U.S. presidents.

Fig. 4.2 Okamoto photograph of former President Eisenhower with Lyndon Johnson

   The remaining two images were both selected by my neighbor as they reminded her of activities in which she became involved in her twenties/thirties; civil rights, anti-war, and women's rights. Although she did mention that there were no images in the collection from which she was presented to select that represented women's rights, the pictures she chose still connected her to the time of her engagement with women's rights.

Image 2

Image 2, shown in Fig. 4.3, in her eyes, represents an era of tumultuous activities and a paradigmatic shift in US history. The photo of Johnson meeting with leaders of the Civil Rights Movement serves as a connection/clue to those racially-charged times and, ultimately, to a shift that made her feel good about the actions of our political leaders.

Image 3

Image 3, shown in Fig. 4.4, Judy feels, gave her a sense of direct witness to the President during the Vietnam War years. A similar sentiment was expressed by former White House Photographer, David Hume Kennerly, during the Ford administration (Estrin, 2013). He stated in an interview about Okamoto's work: "He captured the tension and the fatigue on the face of L.B.J. When you look at the pictures, you just feel like you're there." Judy

Fig. 4.3 Civil rights leaders with President Johnson

Fig. 4.4 Secretary of Defense Robert McNamara and President Johnson

communicated to me later that "…with respect to this photo, I think the look on President Johnson's face tells it all: I'm sure he's thinking WTF….and, at the end of the day, he was correct."

4.1.3   #Archives, #Epidata

To date, we have not been able to locate these three images through an Internet search. Photos apparently taken at these same events are published online; but these specific images which convey different angles, lighting, or content have yet to be found. The mere publication of this provocation makes these retrievable by an archivist working with presidential photos. And this additional documentation may inform archival work of a related image. Or it may lead the archivist to new information. Or it may even lead to the images' owner, who might provide more information or facilitate a connection with the person from which these were acquired. All of these possibilities being potential connections in a web of proximity. Yet none of what is written here directly informs the images (i.e., describes event or what is going on). Nor does our provocation content describe the photo (metadata). Rather what is written here is an appendix, irrelevant (anecdote) or relevant (epidata), to the photos. At the very least, these photos made a stronger connection between Judy and me—a function likely never imagined by Yoichi Okamoto, yet one of value to me.

4.1.4 # Incidental Glance

Laurie was not looking for pictures of presidents; the neighbor's photos were seen as an incidental consequence of performing a neighborly act. Having recently worked on a piece about presidential photographs, she likely had a heightened awareness of such documents, yet there was no intentional seeking at play in the discovery. Seeking understanding about the circumstances of the photos being there led to conversations with the neighbor that went beyond the images and their intended meaning. The photos, in a sense, catalyzed understanding the neighbor. Epidata can be seen as different from data and metadata in its potential for—perhaps even necessity within—unintended stumbling upon a document and the unexpected consequences of the circumstances of that stumbling.

We assert that ultimately usefulness is determined by the user. Users may find documents purposely searching by or formal of idiosyncratic means, or they may stumble upon them in the most unlikely places following a single slender thread of proximity. An intermediary may provide a thread of understanding or suggest a thread for finding. How might a provocation such as the presidential photos on a neighbor's wall probe your thinking? How might it impact your work or personal interests? How might a historian use this information to accomplish their work despite the lack of centrality of information? What other sorts of threads of proximity have you followed? What threads of proximity have you provided to someone?

4.2 Thoughts on Proximity

The tale of Theseus and Ariadne presented us with different facets of proximity. Theseus has to sail from Athens to Knossos just to be proximate with the minotaur. Ariadne becomes close to Theseus and wanting to maintain physical and emotional proximity to him decides to help solve the problem of exiting the labyrinth. Neither Ariadne nor Theseus has any idea of how to escape the labyrinth; however, Ariadne is close to Daedalus, designer of the labyrinth. It is Daedalus who comes up with the idea of using a ball of thread to establish a connection to the exit—a clue, a route marker for regaining proximity to the exit. Had Theseus not met Ariadne and had she not known someone who might have an idea, the Minotaur would no longer have been a threat to Athens, but Theseus would not have returned home.

When you make a grocery list, when you take vacation photos, when you write a journal article you are attempting to put the reader/viewer/user into proximity with you, the author. Grocery lists and vacation photos may be seen by nobody other than you, yet even in these cases, you (the author) are put back in time and, perhaps place—standing before the empty shelf in the pantry two hours earlier, or back at a fjord in Norway in 1973. Damasio asserts that remembering something is putting neural pathways into the same state they were when the target of remembering originally happened. (Damasio, 2010) The documents act as clues, codes to help neural pathways become proximate to some earlier state. In a conversation in 2005, Brian asked Damasio if a novelist or film director could be said to be putting readers and viewers into neural states they had never actually experienced; Damasio replied that he would not argue against that idea.

We posit that meaning is (or is largely) function. Authors generate documents in order to share some functionality—perhaps only with themselves and those close to them or with some wider group; seekers want some functionality. The documents (messages) generated and sought are of all sorts—books, sculptures, movies, symphonies, videos on all manner of topics on social media reading Ginsberg's HOWL, hearing Ginsberg perform HOWL, among myriad others.

4.3   Epidata for *No Place to Go*

Is understanding of an author's coding practices/choices absolutely necessary for functional understanding of a message. We present three provocations to stir discussion. An incomplete understanding of coding practices seems to yield understanding of the message.

This is what one sees on the VIMEO page for Brian's film *No Place to Go*; it is rather like standard bibliographic representations (Fig. 4.5) (O'Connor, 1970). The film was made under the auspices of the federal War on Poverty program in 1970 in Nashua, New Hampshire, where collapse of local industry had led to hard times for many in the city. The half-hour film was shown on local television channels, at U.S. Congressional hearings on funding low-income housing, and was used in training new members of VISTA (Volunteers in Service to America the "domestic Peace Corp.") Several professionals in various fields had gathered statistics, drawn maps, and consulted with government officials; however, there were no photographs, nothing to present the physical evidence of impoverished living. Brian had worked in a companion War on Poverty program and had experience making films, so he volunteered his services.

After a showing of *No Place to Go* in a film theory class at the University of California, Berkeley in 1984, Brian conducted a Q&A session with students and faculty. Among the first comments were praise for the selection of black and white film over color, as shown in Fig. 4.6, and for the use of a sound collage rather than synchronized sound and narration. Brian responded that while he was pleased to hear that those two production decisions

Fig. 4.5 1970—black & white 16 mm film about bad housing, urban renewal, and having no place to go. Made as part of a War on Poverty program in Nashua, New Hampshire

were regarded highly by the audience, he could not take credit for the choice—there was no choice, only necessity.

The film was funded by a local community action group who in turn were funded by federal money from the War on Poverty. Funded is a big word—there was enough money to shoot 45 min of camera original and edit it to half an hour and make a print. The standard for budgeting such a film at the time was to assume it would require shooting at least 50 feet of film for every one foot in the final 1,000-foot, half-hour product. That was generally stated as a "shooting ration of 50:1 shooting ratio; *No Place to Go* had a shooting ratio of 1.5:1. All labor was volunteer.

In 1970 sound was generally not recorded onto the film in the camera—it was recorded on a separate reel to reel tape recorder. To have synchronized sound required some means of assuring the camera and recorder were running at exactly the same speed—no such system was available on such a low-budget production. This constraint actually freed the production team to simply build a collage of images and a collage of sounds. The camera was entirely handheld because Brian wanted to be mobile and did not want to embarrass or overwhelm people in their homes. Black & white film was the only option since the indoor light levels were so low (and the fuses in the homes would not have supported any form of additional lighting). Therefor, black and white film was the only available option. The camera was a government surplus device designed for use in the Arctic (the

Fig. 4.6 Frames from *No Place to Go*

film was shot in the winter); it used a spring motor that had to be wound up before each shot—about 18 s maximum run time per shot. At eight pounds it was considered to be portable.

In comments for a much more recent showing of the film to a film theory course, Brian wrote:

> I should say that I spent a month talking with people in the neighborhood and with city officials and with members of Volunteers in Service to America (VISTA, the domestic Peace Corp) before shooting any film. I asked the residents what they thought we should film; I showed all footage when it came back from the lab to the residents; I asked if there was anything I didn't get right or that was too embarrassed to have go on television. The title comes from a line that many people said to us - this was an urban renewal project to make the city look better and more productive, but there was no plan to re-settle or help the residents of the neighborhood - all of whom were being evicted. The music track by my friend Rob Vergas is a rendition of a piece by Bessie Smith called *No Place to Go* - about a major flood on the Mississippi River in the early 20th century. We got a really cheap guitar from a pawn shop and recorded Rob sitting in the bath tub in my parents' home to get the right sound. The shooting I still like, the editing still works for me, even the rough audio editing works. We were trying to put a human face on a major problem for which there were lots of statistics and surveys but not a lot of emotional buy in.

Here we have a small and intriguing coding/decoding issue: the film sought to achieve an "emotional buy in" not available from tables and charts of figures about the neighborhood. The statistics do not put readers into the same sort of proximity as, for example, the photograph of a boy standing by the door to his derelict home, as shown in Fig. 4.7. City leaders and audiences in later years saw that emotional impact coming in part from technical practices that were necessities. The well-known black and white photographs of the Dust Bowl were largely in black & white and were powerful, though it might be argued that the subjects and compositions and lighting were also significant in the coding. Color would have been Brian's choice—to show the muted, old, faded color palette of poverty in the area at that time; though, in looking back—that is, in engaging with audiences and learning that for some the grey scale palette was a clue, an emotional hook to the burdens of poverty—he might have opted for a mix of color and black and white IF that had been an option.

The message coded with statistics and graphs represented the neighborhood, the people, and the living conditions in a poor neighborhood; yet that message was only marginally meaningful, vaguely functional to many of the people who were making the case for change as well as for those who had to be persuaded. The movie coded the lives of the neighborhood in a way that functioned well for a larger audience, even if some of

Fig. 4.7 Boy at door to his home in *No Place to Go*

the coding was by necessity. The film was successful in promoting legislative changes to aid low-income tenants and Volunteers in Service to America used the film for training new volunteers—that is to say, the film helped bring information to the point of use.

So, we might say the film is the message/document and all of the detail documented here is the epidata that potentially impacts how contemporary audiences see the *No Place to Go*. Much of the detail also addresses the idea that necessity resulted in using black and white film, which not have been Brian's choice, but because of famous historical use black and white photographs to document poverty many viewers unexpectedly applied their antecedent understanding of historical clues to their reading of the film.

4.4   Translation Disease: Proximity Gone Awry?

We recently crafted a conference presentation on translations that don't seem to work; not the early attempts of students or the mistakes of a frantic tourist, but serious productions of crafted words in another language, a novel into film, a piece of music into a different style, an opera set in a different time and place.

We opened with Wilson's assertion: "A translation must preserve the sense of its original ... but there is no imaginable way of saying precisely how much of the sense of the original must be preserved, for a putative translation really to be a translation of some text." (Wilson, 1968)

In the conference proposal we used "disease" in its older sense of "lack of physical comfort, tranquility, state of mind" to describe reactions of original authors and reactions of readers/viewers/listeners of translations. We asserted that when turgid free verse and modern assumptions bury the engaging poetry of Homer, some contemporary readers delight in the majesty of the epic, while others see encrusted layers burying the accessibility of the original—dis-ease. When a high production value music video overwhelms a simple blues song about a tragedy, some audiences delight in that translation, while others see a loss of connection—dis-ease. When a Woody Allen movie is playing with subtitles in another language, some viewers who understand the original sound track are laughing, while others not fluent in English usage in comedy are mystified by the laughter—dis-ease. When a composer sets a two-line Latin poem into a minute and a half portion of a cantata, some listeners hear a lovely musical piece with the same Latin words, while others hear the music burying the raw power of the original—disease.

We apply our web of proximity model to cases of translation that could be argued to bear little resemblance to the original, in order to tease out the role of antecedents in both the production and perception of translation documents. Translation at its best is fraught with difficulties; "poor" translations can shroud, transform, even completely obliterate the author's intended meaning. An audience might like a lovely production that actually obscures the original or might find in it an entry into exploring the original. We do not claim to have a way of saying how much of the sense of the original ought to be preserved or in what way the preservation should happen; however, our model will provoke some ways of thinking about the problems of preserving some sense of the original. That is to say, our notions of proximity and epidata—especially considerations of how antecedents effect expression of the data of the message and the resulting interpretation. This is not an attempt to lay a critical judgment on any particular translation, nor is it an attempt to join the fray of argument over the place of authorial intent in "proper" understanding; rather it is an attempt to afford another means of bringing antecedents of all participants in the conversation into play.

4.5   How Many More Lucias Are Out There?

Revisiting our early example of Prometheus and Ariandne where a ball of yarn serves as a clue, we saw that epidata helped solve Prometheus true challenge of exiting the labyrinth successfully. We might consider this a microcosmic approach where epidata connects a single person to a single source of information and then to a functional resolution to a problem. Anytime information successfully solves an issue makes for a valuable information use.

But what of the power of epidata on a larger scale. Historically, opera was performed for the wealthy and noble classes. With the opening of the Teatro San Cassiano in Venice in 1637, opera became accessible to the general public. At this juncture, opera subdivided into two camps: opera seria and opera buffa. Opera seria focused on storylines around myth and kings with nobility remaining as the intended audience. Typically performed in three acts, the duration tested many audiences. On the other hand, opera buffa was comedic in nature portraying everyday stories. The air of these operas was plainer and presented in two acts making it attractive to a wider audience. Audiences to opera require processing a multiplicity of stimuli including the following of the plot and observations of sight, sounds, props, and stage design orchestrating a world that mirrors reality. The productions are a quintessential blurring of reality and fiction.

Since the opening of the Teatro San Cassiano, more people have had access to enjoy opera. Yet opera remained elusive to the masses due to limited access to opera house venues. Today, the most notable opera house is the Metropolitan Opera in New York City, which is a far cry from medium-sized cities and certainly not accessible to small town America. In an effort to build revenues and attract new audiences, the Metropolitan Opera (The Met) debuted the Met broadcast series in December 2006 bringing opera to communities around the globe via satellite into movie theaters. Since 2006 the Met has expanded its Live in HD performances covering more productions and broadcast to a growing number of cinemas throughout the world. And performances are now available through On-demand subscription to a TV near you. Yet there remains the issue of opera

closely reflecting our world. The static setting of the opera classics (1600–1900s) eventu-
ally separated them from mirroring reality, distancing understanding for a mass audience
immersed in a modern world. Proximity was stretched.

Enter the creative complex of opera production where, recently, Australian director
Simon Stone translated the classic *Lucia di Lammermoor*. This opera tragico, written in
three acts, is clearly intended for noble and wealthy audiences of the nineteenth century.
Stone envisioned the power of bringing to the story a contemporary flair that would
resonate with audiences today. Donizetti's classic opera was composed in 1835 at a time
when male honor and dignity were at the center of real life. Working with the original
score and libretto (synchronic data), Stone reworked the sight, sound, props, and stage
design to reframe the story metropolitan Detroit in the 2000 (diachronicity). Stone saw
reworking the storyline as a "great opportunity to participate in that conversation and
confront the simultaneous absurdity and danger of the idea that a man's pride is more
important than a woman's life."

Let's revisit epidata and its origins to get a better look at it playing out in Stone's recre-
ation of *Lucia di Lammermoor* affording a more likely connection for today's audience.
We conceived epidata from epigenetics. Defined as the study of changes in organisms
caused by modification of gene expression rather than alteration of the genetic code itself.
epigenetics makes changes to the genes, small or large, to modify how the original gene
is expressed. Turning to information, we define epidata as changes in content around or
accompanying information that modifies meaning or understanding rather than alteration
of the information itself.

The score and libretto for *Lucia di Lammermoor* are just squiggles on a page, as seen
in Fig. 4.8, until it is performed by an orchestra and vocalists (Donizetti and Cammarano,
1992). Stone left Donizetti's original music composition intact. He made changes to the
setting through alterations of visual presentation. Through this translation to a modern-
day tragedy where opioids impact human behaviors and outcomes. In Stone's production,
he conveys a world where what life promises in youth falls short in adulthood as the
world evolves, leaving one feeling betrayed. "…, anyone else's feelings of betrayal or
abuse can't compete with his own sense of loss from the position he lived in earlier to
where he's landed. So, you have capitalist systems, which only care about profit, abusing
workers, who in turn become perpetrators of abuse against other members of society,
especially women. It's a sad, vicious cycle." (Dobkin & Goodwin, 2022)

In both the original opera production and the 2022 Met Opera production of *Lucia di
Lammermoor*, the setting conveys a grim air. True to both versions, the characters hold
to the hope of finding happiness despite life conditions. Both opera productions convey
an underpinning of a fading dream. Stone shares "To me, that's the way Donizetti's score
sounds, as well. He creates a sense of nostalgia from the get-go—the music of fading
dreams." (Dobkin & Goodwin, 2022)

Fig. 4.8 *Lucia di Lammermoor* score and libretto, p. 15

In review, epidata is a phenomenon that has been occurring all around us. Outside of the context of formal information practice (i.e., reference desk, etc.), we see the impetus of influence played out in the information around information. The original information (like Donizetti's music score) remains unchanged. It is the surrounding information that is reorchestrated to impact meaning and adoption of information. For example, in the presidential photos of Johnson's emergency swearing-in ceremony, what if the blood on Jackie Kennedy's dress was removed? Would this change the meaning for current day viewers of the photo? If pictures of the outside of Airforce One on that day were shared alongside of the photo, would the meaning change? What if we placed all the same people, in the same staging, in the Oval Office, would the meaning change?

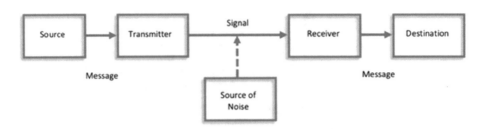

Fig. 4.9 Typical graphical representation of Shannon's communication model

4.6   Thinkering[1] with Shannon

Here we have in Fig. 4.9 a typical graphical representation of Shannon's linear model of communication. For our purposes, we wish to relabel/reorder some of the elements while staying true to Shannon's construct. We want to be talking about authored messages. Ordinarily these would be books, articles, videos, streaming audio files, grocery lists, and the like. Shannon, of course, cleaved meaning from message, concerning himself only with the squiggles, dots, or bits arriving at their destination just as they had begun at their source. Meaning cannot be derived without the message but it is not inherent in the message.

We have played with layering meaning on top of the message map, as represented in Fig. 4.10, as a means of clarifying our notion of proximity. What if we add a layer of author and user—the author coding some medium with the intention of preserving/sending some meaning; the recipient decoding the message for meaning to resolve some issue? What if we then add a layer of all the antecedents stimulating the author and all the antecedents stimulating the recipient of the message? We added the notion of "antecedents" to account for all the stimuli on the author's constructing the message: e.g., purpose, intended audience, language, facility with the chosen medium. We do the same for the seeker to account for similar stimuli. This might be seen as similar to the notion of "common ground," in which the probability of the author's intended meaning being what is decoded by the seeker increases as the number of antecedents in common increases. This does not necessarily mean that seekers do not/cannot use messages for quite different purposes than those intended by the author.

---

[1] The term is derived by Wyatt and O'Connor from Marshall McLuhan's work and think by tinkering. See O'Connor, B., & Wyatt, R. B. (2004). *Photo provocations: Thinkink in, with and about potographs.* Lanham, Maryland: Scarerow Press. P. 12.

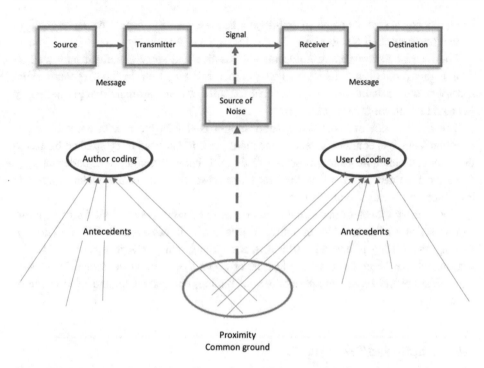

Fig. 4.10 Antecedents layered over Shannon's model for a model of proximity consistent with Shannon

Can we say the functionality of a message that arrives just as it was sent is dependent on the degree of overlap between the antecedents shaping the author's coding abilities and the antecedents shaping the recipient's decoding abilities? That is, the functionality is influenced by the proximity that holds between the partners in the conversation or that which can be provided as epidata.

We might say that second-hand knowledge—books, articles, photographs, YouTube videos, Twitter, and all the progeny—provides a way of being close to events we did not witness, to experts whom we do not know, places we have not been. Second-hand knowledge helps us achieve a certain nearness, an approximation of first-hand knowledge. The common ground an author and a recipient/seeker do not have can be seen as noise.

4.7   Representation

By definition, representation throws out some of the attributes of that which is being represented. This is often just what one wants. If this were not the case, then we would also have to examine entire documents as the first stage of a search; this would be a significant enlargement of search space and require inordinate amounts of search time.

How, though, are we to come to solutions when the question cannot be represented in system terms or at all a priori.

The little hook between an unformulated question and some document may well be what is tossed out during the representation process. It may even be the case that a satisfactory document is in hand but is rejected for want of some attribute thought necessary yet could be shown to be unnecessary.

The achievement of "common ground" might well take place with common topic, common language, common level of education; yet it might need to be sparked by something else in common. Here we do not distinguish between finding and understanding. Having a document in hand that does not make sense, does not enable understanding, is not functional is as good as not found.

The primary thrusts of our discussion are that the goal or result of all forms of information seeking is be close to an authority with useful information; that a good deal of seeking and finding is beyond retrieval; that anything that stands in the way of useful information can be modeled as interference or noise; that noise is situational; and, especially, that epidata expand the possibilities of finding, understanding, and making use of documents.

4.8 Epilogue/Prologue

When we first began to write, we used the metaphor of the tree falling in the woods, as pictured in Fig. 4.11, with nobody there to hear it—in order to step away from the concept of organizational, pre-coordinate representation of documents. We asked: If a tree falls in a forest and no one is around to hear it, does it make a sound? This philosophical thought experiment has probed minds since the early 1700s (Berkeley and Turbayne, 1957). Responses to the question emerged from the physical science of sound where the predominant thinking is that sound is the sensation excited in the ear when air or other medium is set in motion. The major premise in this proposition is that there must be a receiver (hearer) in order for sound to be present. More technically speaking, sound is only recognized specifically at nerve centers further emphasizing the role of the receiver to justify existence of sound (*Scientific American*, 1884). Simply put, no ears, no sound…tree falls in silence.

Why open with this philosophical provocation? Sound depends on some form of proximity. We want to use this metaphor as a probe into notions of proximity in the woods and forests of the information environment. Forests and woodlands, like libraries and the World Wide Web, have long been places useful to humans (Fig. 4.12). Jaegels uses the term "close associations" and "direct connections" to describe the relationships our forebears had with forests: [these] gave boatbuilders and other craftsmen of the past a kind of intimacy that fostered communal knowledge." (Jaegles, 2021) This sort of intimacy is at the heart of our argument and our model of proximity.

Fig. 4.11 Tree fallen in New York woods

We wanted, among other things, to convey the idea that much of epidata is unpredictable—ought we to include in the bibliographic record for each work a caution that in "x" years anyone who comes across the bibliographic record ought to check with the publisher's style manual at the time of publication for policies on gender neutral pronouns? How would we even predict what elements of a culture might change just enough so that there could be a "slight" but consequential misunderstanding? Our concept of proximity and its utility hinge on fostering the sensitivity to "close associations" and "direct connections" enabled by epidata.

4.9 Provocations

The preceding pages represent our thinking about proximity and epidata along with provocations we have used to moor our thinking. Here we add a few more provocations to seed our ongoing conversations

What sort of thread would link a seeker to a functional message?

Can we see documents as conversants between situations?

Fig. 4.12 Boat builders' understanding of trees, engraving in Panckoucke's *Encylopedie Methodique: Marine*, 1783

What if finding and understanding/functionality are not always purposeful/intended? Are we in some way addressing Wilson's assertion about accessibility?

> All the documents in the library are immediately available to me, but they are not all accessible to me.

Wilson lists linguistic inaccessibility, conceptual inaccessibility, and critical inaccessibility (Wilson, 1977). Can we argue that notions of epidata and proximity open up awareness of subtleties within each of these constructs of inaccessibility that might well go undetected, yet render a work functionally inaccessible without the user realizing it.

When someone is seeking documental help to solve a problem, make a decision, perform a function, or better understand something, they are attempting to be close to an authority, they are seeking proximity to firsthand knowledge. In common systems of document retrieval, whether pre-coordinate or post coordinate, the assumption is that a representation of a question state and some representations of documents run through a similarity engine will yield a document or set of documents equal to the seeker's task.

What about those situations in which documents have not yet been trieved or the meaning is unclear? The representations of the documents may well be constructed "properly" while the seeker is unclear about the question or while the seeker does not even realize there is a question.

4.10   Invitation

Epidata is the term we use for links between the antecedents of authorship and those of seeking to see how we might aid in the looking for the attributes, the hooks, that might yield finding and understanding yet are not part of the data and metadata ordinarily used to guide seekers. The spectrum of data, metadata, epidata is continuous; the utility of any portion of the spectrum to a seeker depends on the quest.

What could count as epidata? How might one know when epidata would be useful? Do digital platforms enable virtual proximity?

Can we say, then, that representing documents with subject headings or strings of key words or similar extra-verbal representations functions for bringing seekers and documents together for certain sorts of questions and does quite the opposite for certain sorts of questions. They can bring seekers into proximity with firsthand knowledge in some instances, yet stand in the way of proximity for others. Might it be that the smallest or least likely attributes of a seeker/document couplet will be of great use, precisely because the "obvious" attributes are of little use.

Can we model the conversation/dance between author and seeker as the co-authorship of functionality?

Can we say that proximity is polydirectional; common ground is polydirectional.

Can we say that a focus on retrieval not enough? Since re-trieve means "to find again," it implies that what is needed has already been found.

What are we saying? Functionality of a message hinges on more than data and metadata.

What are we not saying? We are not saying we have a complete model; we have a metaphor to organize observations and thoughts and conversations.

What are we not seeing? We ask you: what are we missing? We invite your input. Bring your clue and join the continuing conversation by email and on our forthcoming podcast.

Laurie & Brian

References

Berkeley, G., & Turbayne, C. M. (1957). *George Berkeley: A treatise concerning the principles of human knowledge*. Bobbs Merrill Co.

Bonnici, L. J., & O'Connor, B. C. (2021). More than meets the eye: Proximity to crises through presidential photographs. *Proceedings from the Document Academy, 8*(2), Article 14. https://doi.org/10.35492/docam/8/2/14. Available at: https://ideaexchange.uakron.edu/docam/vol8/iss2/14

Damasio, A. (2010). *Self comes to mind: Constructing the conscious brain*. Vintage Books.

Dobkin, M., & Goodwin, J. (2022). *Fading dreams*. The Metropolitan Opera. https://www.metopera.org/discover/articles/fading-dreams/

Donizetti, G., & Cammarano, S. (1992). *Lucia di Lammermoor: In full score*. International Music Score Library Project (IMSLP). https://s9.imslp.org/files/imglnks/usimg/5/5f/IMSLP111609-PMLP51145-Donizetti_-_Lucia_di_Lammermoor_-_Act_I_(orch._score).pdf

Estrin, J. (2013). Photographing the White House from the Inside. Lens blog, The New York Times. https://lens.blogs.nytimes.com/2013/12/10/photographing-the-white-house-from-the-inside/

Jaegles, R. (2021). All that we may not see. *WoodenBoat, 279*, 88–89.

O'Connor, B. (1970). *No Place to Go*. Motion picture. Available at: https://vimeo.com/257826808

Scientific American, April 15, 1884 (p. 218)

Wilson, P. (1968). *Two kinds of power: An essay on bibliographical control*. University of California Press.

Wilson, P. (1977). *Public knowledge, private ignorance: Toward a library and information policy*. Greenwood.

Printed in the United States
by Baker & Taylor Publisher Services